5G-Enabled Technology for Smart City and Urbanization System

This book examines the applications, trends and challenges of 5G-Enabled technologies for Smart City and Urbanization systems. It addresses the challenges of bringing such capabilities of 5G-enabled technologies for smart cities and urbanization into practice by presenting the theoretical as well as technical research outcomes with case studies. It covers key areas, including smart building, smart health care, smart mobility, smart living, smart surveillance, and IOT-based systems. It explains how these systems are connected using different technologies that support 5G access and control protocols.

- Offers a comprehensive understanding of the emergence of 5G technology and its integration with IoT, Big Data and Artificial Intelligence for Smart City and Urbanization.
- Focuses on useful applications of smart city and urbanization, which can enhance different aspects of urban life.
- Explores the advantages of using massive IoT and predictive analytics approaches in smart cities.
- IoT, Big Data, Deep learning and machine learning techniques are explained to fuel smart cities and urbanization system.
- Addresses both theoretical and technical research outcomes related to smart city and urbanization with 5G technology.

It serves as a valuable reference for graduate students, researchers, and many practitioners seeking to deepen their knowledge and engage with the latest advancements in the areas of Smart Cities and Urbanization systems.

5G-Enabled Technology for Smart City and Urbanization System

Edited by
Susheela Hooda
Vidhu Kiran
Rupali Gill
Durgesh Srivastava
Jabar H. Yousif

CRC Press
Taylor & Francis Group
Boca Raton London New York

CRC Press is an imprint of the
Taylor & Francis Group, an **informa** business

A CHAPMAN & HALL BOOK

Designed cover image: ShutterStock

First edition published 2025
by CRC Press
2385 NW Executive Center Drive, Suite 320, Boca Raton FL 33431

and by CRC Press
4 Park Square, Milton Park, Abingdon, Oxon, OX14 4RN

CRC Press is an imprint of Taylor & Francis Group, LLC

Library of Congress Cataloging-in-Publication Data
Names: Hooda, Susheela, editor. | Kiran, Vidhu, editor. | Gill, Rupali, editor. | Srivastava, Durgesh (Computer scientist), editor. | Yousif, Jabar H., editor.
Title: 5G-enabled technology for smart city and urbanization system / [edited by] Susheela Hooda, Vidhu Kiran, Rupali Gill, Durgesh Srivastava, Jabar H. Yousif.
Description: First edition. | Boca Raton : C&H/CRC Press, 2025. | Includes bibliographical references and index. | Summary: "This book examines the applications, trends and challenges of 5G-Enabled technologies for Smart City and Urbanization systems.
Identifiers: LCCN 2024028716 (print) | LCCN 2024028717 (ebook) |
ISBN 9781032740133 (hbk) | ISBN 9781032741567 (pbk) | ISBN 9781003467892 (ebk)
Subjects: LCSH: Smart cities--Technological innovations. | 5G mobile communication systems--Technological innovations.
Classification: LCC TD159.4 .A124 2025 (print) | LCC TD159.4 (ebook) |
DDC 307.1/416--dc23/eng/20240729
LC record available at https://lccn.loc.gov/2024028716
LC ebook record available at https://lccn.loc.gov/2024028717

ISBN: 978-1-032-74013-3 (hbk)
ISBN: 978-1-032-74156-7 (pbk)
ISBN: 978-1-003-46789-2 (ebk)

DOI: 10.1201/9781003467892

Typeset in Times
by SPi Technologies India Pvt Ltd (Straive)

Contents

About Editors

Dr. Susheela Hooda is an Associate Professor, at Chitkara University Institute of Engineering & Technology, Chitkara University, Rajpura, Punjab, India since 2021. She completed her Ph.D. in Computer Science & Applications in 2018, from Maharshi Dayanand University, Rohtak, Haryana India. She has more than 15 years of teaching and research experience. Dr. Susheela is a distinguished researcher in the areas of software engineering, aspect-oriented software development, software testing, cloud computing, Artificial Intelligence, Machine Learning. She has published more than thirty technical research papers in leading journals and conferences. As per editorial experience is concerned, she is serving as an editor for a renowned publisher. She has published and granted more than 10 national and international patents.

Dr. Vidhu Kiran Sharma is an Assistant Professor in the Department of Computer Science and Engineering, CDLSIET, Haryana, India since 2023. She completed her Ph.D. in Computer Science & Engineering from Maharaja Ranjit Singh State Technical University, Punjab, India. Dr. Vidhu is a renowned researcher in the areas of Internet of Things, Wireless Sensor Networks, Security, Blockchain, Cryptography, UAV, Machine Learning and Software Engineering. She has published more than thirty technical research papers in leading journals and conferences. She published 5 International and national patents.

Dr. Rupali Gill is an Associate Professor and Dean at Chitkara University Institute of Engineering & Technology, Chitkara University, Rajpura, Punjab, India. She completed her Ph.D. in Computer Science and Engineering in 2021, from Chitkara University, Rajpura, Punjab, India. Dr. Rupali Gill is a distinguished researcher in the areas of Image Processing, Cloud Computing, Artificial Intelligence, Machine Learning. She has published more than 40 technical research papers in leading journals and conferences. She has organized International Conferences and has been granted a published patent.

Dr. Durgesh Srivastava is an Associate Professor, at Chitkara University Institute of Engineering & Technology, Chitkara University, Rajpura, Punjab, India. He received his PhD degree in Computer Science & Engineering from IKG Punjab Technical University, Jalandhar, Punjab, India in 2020. He received B. Tech. degree in Information & Technology (IT) from MIET, Meerut, UP in 2006 and ME in Software Engineering from Birla Institute of Technology (BIT), Mesra, Ranchi, Jharkhand, India in 2008. He has 14 years of research and academic experience. His research interests include Machine Learning, Soft Computing, Pattern Recognition, Software engineering, modelling, design, etc. He has published more than 30 papers in reputed international/national journals and conferences/seminars. As per editorial experience is concerned, he is serving as an editor for CRC Press (Taylor & Francis). He has published several patents and copyrights in the field of computer software. He has organized and attended various workshops during his teaching career. He is a life member of a different professional society.

Dr. Jabar H. Yousif Associate Prof. at Faculty of Computing and Information Technology, Sohar University, Oman. Ph.D. in Information Science & Technology, M.Sc. & B.Sc. in Computer Science. Postdoctoral fellowship in Virtual Reality. More than 25 years of teaching experience. He has published more than 90 papers & books in Artificial Intelligent, Cloud Computing, Soft Computing, Artificial Neural Networks, Natural Language Processing, Arabic text Processing & Virtual Reality. Editorial board & reviewer for many scientific journals and conferences.

Contributors

Abdallah Abualkishik
Faculty of Computing and Information
 Technology, Sohar University
Sohar, Oman

Khaled Abuhmaidan
Faculty of Computing and Information
 Technology, Sohar University
Sohar, Oman

Suhaib Ahmed
Department of Electronics and
 Communication Engineering,
 Model Institute of Engineering and
 Technology
Jammu (J&K), India

Mahmood Al-Bahri
Faculty of Computing and IT, Sohar
 University
Sultanate of Oman

Ghassan Al-Kindi
Faculty of Engineering, Sohar University
Sohar, Oman

Wasin Al Kishri
Faculty of computer studies, Arab Open
 University Sultanate of Oman
Oman

Marwan Alshar'e
Faculty of Computing and Information
 Technology, Sohar University
Sohar, Oman

YPS Berwal
Ch. Devi Lal State Institute of
 Engineering & Technology,
 Panniwala Mota, Sirsa
Haryana, India

Udaibir Singh Bhathal
Chitkara University Institute of
 Engineering & Technology, Chitkara
 University
Punjab, India

Garima Chopra
Chitkara University Institute of
 Engineering & Technology, Chitkara
 University
Punjab, India

Sanjay Dahiya
Ch. Devi Lal State Institute of
 Engineering & Technology
Panniwala Mota, Sirsa, Haryana,
 India

Pummy Dhiman
Chitkara University Institute of
 Engineering & Technology, Chitkara
 University
Punjab, India

Rupali Gill
Chitkara University Institute of
 Engineering & Technology, Chitkara
 University
Punjab, India

Monisha Gupta
Computer Science and Engineering,
 Presidency University
Bangalore, India

Priyanka Gupta
Chitkara University Institute of
 Engineering & Technology, Chitkara
 University
Punjab, India

Susheela Hooda
Chitkara University Institute of
    Engineering & Technology, Chitkara
    University
Punjab, India

Bikram Kar
Sanaka Educational Trusts Group of
    Institutions, Durgapur
West Bengal, India

Rashmeen Kaur
Chitkara University Institute of
    Engineering & Technology, Chitkara
    University
Punjab, India

Gagandeep Kaur
Chitkara University Institute of
    Engineering & Technology, Chitkara
    University
Punjab, India

Gaganpreet Kaur
Chitkara University Institute of
    Engineering & Technology, Chitkara
    University
Punjab, India

Mandeep Kaur
Chitkara University Institute of
    Engineering & Technology, Chitkara
    University
Punjab, India

Rupinder Kaur
Ch. Devi Lal State Institute of
    Engineering & Technology,
    Panniwala Mota, Sirsa
Haryana, India

Md. Sameeruddin Khan
Computer Science and Engineering &
    Information Science, Presidency
    University
Bangalore, India

Sameeksha Khare
Manav Rachna International Institute of
    Research and Studies
Faridabad, India

Ahmad Kayed
Computing and Information
    Technology, Sohar University
Sohar, Oman

Amit Kumar
Sanaka Educational Trusts Group of
    Institutions, Durgapur
West Bengal, India

Kamal Kumar
Ch. Devi Lal State Institute of
    Engineering & Technology,
    Panniwala Mota, Sirsa
Haryana, India

Manoj Kumar Mahto
Vignan Institute of Technology and
    Science, Deshmukhi, Hyderabad
Telangana, India

Srinivas Mishra
Computer Science and Engineering,
    Presidency University
Bangalore, India

Ala Odeibat
Computing and Information
    Technology, Sohar University
Sohar, Oman

Poonam
Ch. Devi Lal State Institute of
    Engineering & Technology,
    Panniwala Mota, Sirsa
Haryana, India

Pratibha
Chitkara University Institute of
    Engineering & Technology, Chitkara
    University
Punjab, India

Varsha Rani
Ch. Devi Lal State Institute of
  Engineering & Technology,
  Panniwala Mota, Sirsa
Haryana, India

Rashi Sahay
Manav Rachna International Institute of
  Research and Studies
Faridabad, India

Preeti Sharma
Chitkara University Institute of
  Engineering & Technology, Chitkara
  University
Punjab, India

Shubham Sharma
Chitkara University Institute of
  Engineering & Technology, Chitkara
  University
Punjab, India

Vidhu Kiran Sharma
Ch. Devi Lal State Institute of
  Engineering & Technology,
  Panniwala Mota, Sirsa
Haryana, India

Anisha Singh
Chitkara University Institute of
  Engineering & Technology, Chitkara
  University
Punjab, India

Mahender Singh Kaswan
School of Mechanical Engineering,
  Lovely Professional University
India

Sukhraj Singh
Chitkara University Institute of
  Engineering & Technology, Chitkara
  University
Punjab, India

Durgesh K. Srivastava
Chitkara University Institute of
  Engineering & Technology, Chitkara
  University
Punjab, India

Righa Tandon
Chitkara University Institute of
  Engineering & Technology, Chitkara
  University
Punjab, India

Jabar H. Yousif
Computing and Information
  Technology, Sohar University
Sohar, Oman

1 5G with Emerging Technology and Its Implication

Vidhu Kiran Sharma
Ch. Devi Lal State Institute of Engineering & Technology,
Sirsa, India

Susheela Hooda and Rupali Gill
Chitkara University Institute of Engineering and Technology,
Chitkara University, Punjab, India

YPS Berwal and Sanjay Dahiya
Ch. Devi Lal State Institute of Engineering & Technology,
Sirsa, India

1.1 INTRODUCTION

Over the past few decades, mobile technology has advanced at a rapid pace, with new standards and capabilities being introduced in short succession. In 1980, first generation of mobile networks came into existence to provide basic voice services using analogue transmission. 2G networks, launched in the 1990s, made the transition to digital networks and introduced services like text messaging. 3G brought mobile internet and broadband capabilities in the early 2000s (Gupta et al. 2023). As of 2020, most networks around the world have transitioned to 4G LTE networks, providing faster connection speeds and supporting advanced applications and services. Now, the industry is looking ahead to the next major phase: 5G networks. 5G promises significant improvements in speed, responsiveness, and capacity compared to prior generations. Network operators around the world have begun rolling out early 5G networks, though widespread adoption is still on the horizon. As we look to this future, it's helpful to understand the key drivers pushing mobile technology forward and what specific benefits 5G aims to provide (Zhang 2019).

There are a few key factors motivating the development and adoption of 5G networks:

- Need for Speed
  Each successive generation has brought impressive leaps in connection speeds. Peak 4G LTE speeds can reach around 1 Gbps, though average user speeds tend to be 50–100 Mbps. 5G networks are targeting peak speeds up

to 20 Gbps, with average user experiences of 100–900 Mbps. These major speed boosts will support uses in ultra-high-definition video streaming, augmented and virtual reality apps, real-time gaming, and more advanced mobile applications (Rodriguez 2015).

- Connecting Everything
  Prior mobile networks were built primarily to connect individual human users. 5G adopts a more expansive vision of connecting not just smartphones and computers, but enabling machine-to-machine communication across a wide range of devices, sensors, and objects. The goal is to build an Internet of Things (IoT) ecosystem powered by fast, reliable cellular connectivity. Connected vehicles, smart infrastructure, industrial automation, wearables, and home appliances are all candidates for adoption (Shetty 2021).
- Low Latency
  In addition to higher overall speeds, 5G aims to provide extremely low latency. Latency refers to the time it takes for devices to initiate a communication or data transfer to the network and receive a response. 4G networks have latencies of around 50–100 milliseconds. 5G is targeting 1-millisecond latency for uses like autonomous vehicles that require an instant connection. Lower latencies will drive more real-time interactivity in applications.
- Network Reliability
  Mission-critical applications like remote surgery, power grids, and industrial automation cannot afford network interruptions or performance degradations. While no network offers perfect reliability, 5G introduces new network architectures, communication protocols, and infrastructure densification to offer "five 9s availability" – 99.999% uptime. For systems that demand continuous connectivity, 5G offers reliability on par with fibre optic networks (García et al. 2020).

1.1.1 5G DEVELOPMENT

There is an extensive ecosystem of organizations involved in researching, standardizing, and deploying 5G networks globally:

- Telecommunication carriers are the core investors in 5G networks, attracted by opportunities to increase revenues and remain competitive in the mobile connectivity market. Top global carriers like Verizon, AT&T, China Mobile, and SK Telecom are leading investments measured in billions of dollars.
- Cell phone manufacturers are developing 5G-compatible devices to bring the benefits of new networks to consumers. Samsung, LG, Apple, Xiaomi, and others aim to be first to market as 5G adoption spreads.
- Chip designers like Qualcomm and Intel are creating advanced semiconductors, modems, and other hardware integral to powering 5G connectivity in phones and infrastructure. Huawei and ZTE are also key hardware suppliers.
- 3GPP industry consortium has representatives from hundreds of member companies defining the technical 5G specification. Interoperability is critical.

- National governments view leadership in 5G as conferring economic and strategic advantages. Government initiatives aim to spur investments in research and adoption.
- Universities bring academic research expertise, exploring innovations in spectrum usage, network architectures, hardware capabilities, and applications.
- With collaboration across these stakeholders, widespread 5G availability could be achieved within the next 5 years across most developed countries (Pujari et al. 2021).

1.1.2 CAPABILITIES OF 5G NETWORKS

5G standard defines a complex array of features and connectivity modes to address diverse use cases and deployment scenarios. At a high level, the specification categorizes 5G services into three classes:

1.1.2.1 Enhanced Mobile Broadband (eMBB)

The eMBB category focuses on empowering consumer mobile internet and entertainment applications with significant performance boosts:

- User experienced speed of 100–900 Mbps.
- Less than 5 milliseconds of air interface latency.
- 99.999% availability and 100% coverage for seamless connectivity.
- 60 GHz and higher frequency Wave spectrum usage.
- Up to 100× bandwidth per unit area and 10× spectral efficiency over 4G.

With eMBB, 5G brings fibre-like speeds to mobile users, supporting ultra-high-definition video, virtual reality immersion, rich multiplayer gaming, and other advanced experiences. eMBB establishes 5G as the new standard for mobile broadband connectivity (Khuntia et al. 2021).

1.1.2.2 Massive Machine-Type Communications (mMTC)

The mMTC service category connects IoT ecosystems comprised of millions of low-power and low-cost devices that need to efficiently send small amounts of non-continuous data. mMTC enables applications like:

- Smart cities with connected infrastructure, meters, sensors, cameras, and vehicles.
- Logistics tracking across production, shipping, and retail locations.
- Health devices for patient monitoring, fitness tracking, and elderly assistance.
- Smart homes with appliances, lighting, heating/AC, security, and entertainment.

mMTC focuses on supporting massive device scale, extended coverage to hard-to-reach areas, ultra power efficiency, and multi-year battery life for simple devices aimed at depositing small data payloads in a network.

1.1.2.3   Ultra-Reliable Low Latency Communication (URLLC)

URLLC brings guaranteed low-latency transmission and strong reliability assurances for mission-critical control and automation use cases including:

- Industrial automation with remotely controlled robotics and assembly lines.
- Autonomous vehicles requiring instantaneous data for navigation and obstacle avoidance.
- Smart power grids that rely on real-time monitoring and control.
- Remote medical procedures demanding instant transmission of HD video and sensations for immersive realism.
- More responsive augmented and virtual reality experiences.
- Here latency refers not just to air interface delays of ~1 millisecond but also guaranteed uplink-downlink latencies of less than 5 milliseconds including transmission processing. Reliability specifies packet error rates down to 10^{-9} for crucial messages and links availability exceeding 99.999% uptime (Pons et al. 2023).

1.1.3   TECHNICAL APPROACHES TO ACHIEVING 5G GOALS

Delivering transformative 5G services requires rethinking existing mobile network architectures using new technologies and infrastructure strategies:

- Millimetre wave radio and beam forming – Extreme bandwidth in higher 24 GHz to 100 GHz frequencies demands advanced radio capabilities for tracking best signal paths. These signals only travel short distances and are blocked by objects.
- Small cell densification – Higher frequency signals require cells placed every 100–250 meters in urban areas with interspersed coverage fillers. 5G may deploy over 10× as many base stations as prior gens.
- Massive MIMO antennas – Large antenna arrays with 128–256 elements for more precise transmission across many spatial channels. Advanced signal processing optimizes downlinks to many users simultaneously.
- Dynamic spectrum sharing – Opportunistic usage of any unused spectrum across licensed, shared and unlicensed bands depending on availability and performance needs.
- Next-gen core and edge cloud infrastructure – Finely distributed network and computing capabilities to enable real-time responsiveness, scalability and flexibility. Reduces distance data traverses.
- Network slicing management – Logically partitioning network to optimize performance, efficiency and costs per use case or customer vertical. Enables differentiated services (Jyoti et al. 2011).
- Smart connected machines and sensors – Billions of low-power devices with programmable connectivity and intelligence expanding human-digital interfaces.
- Together these and other key innovations aim to deliver the disruptive 5G vision and support continued mobile technology leadership over the next decade (Loncar et al. 2019).

1.1.4 Impact of 5G on Consumers And Businesses

For the average consumer, 5G networks promise to power new mobile experiences allowing people to stream, download, game, chat, shop and browse on the go at fibre-like speeds with near-instant responsiveness. From data-hungry video buffs to casual web surfers, 5G offers everyone faster performance when away from Wi-Fi. When 5G coverage is paired with advancements on the device side – folding touch-screens, 3D sensors, personal AI assistance, and expanded camera capabilities – we should expect groundbreaking phone capabilities leading users to upgrade devices. However, the impacts of 5G extend far beyond the realm of personal mobile connectivity. Across nearly every industry, 5G has the power to accelerate disruption and opportunity:

- Manufacturing – Real-time automation, tracking, monitoring and control of processes and equipment enables smart factories with 30% production gains. Human workers supported by assisted reality tools.
- Healthcare – Wearable continuously transmit patient vitals for remote monitoring and AI diagnosis. Robots assist surgeries based on 5G video and instrument telemetry. Routine visits conducted via holograms. Genomics revolutionized by analysing massive datasets.
- Transportation – Interconnected vehicles, traffic signals, cameras, public transit and smart highways cooperate to route flow optimally. Self-driving delivery trucks gaining rapid adoption. Drones supported.
- Energy – Distributed 5G-enabled sensors help balance loads, isolate faults, optimize production mixes, prevent outages and drive resource efficiency to cut waste by 40% with perfect availability.
- Media – Ubiquitous high-fidelity streaming of live events and VR experiences with customized points of view. Broadcast innovations create more personalized, interactive content (Blanco et al. 2017).

1.1.5 Advantages of 5G Technology

- Large bandwidth two-way shaping.
- More proactive and effective.

1.1.6 Disadvantages of 5G Technology

While 5G technology aims to resolve all radio signal issues and mobile connectivity challenges, some limitations exist currently in aspects like security, privacy and lack of ubiquitous deployment (Chettri et al. 2019).

- The technology remains unclear on processes and capabilities. Research is still ongoing.
- Achieving high speeds is difficult presently since technology support is unavailable in most global regions.
- Legacy devices cannot access 5G networks. Replacing all devices is expensive for businesses.

- Infrastructure development requires massive investments.
- Security and privacy concerns need resolutions.

1.1.7 FUTURE OF 5G TECHNOLOGY

Future 5G networks promise enhanced performance, quality of service, lower latency, higher bandwidth and greatly improved user experiences for consumers and businesses, including use cases like telemedicine and cloud gaming. According to Sergey Seletskyi, Senior Solutions Architect at Intellias, the Internet of Things (IoT) will transform due to 5G advancements. However, most industries worldwide will not be covered for years. While most people will connect wirelessly to 5G, Wi-Fi will likely handle local wireless connectivity initially. But eventually, there may be a single converged wireless standard as 5G makes Wi-Fi redundant, given its ubiquity today. As more consumers demand improved spectrum, capacity and speeds, the early 5G deployments in many countries utilize sub-6 GHz frequencies comparable to existing mobile and Wi-Fi networks.

1.2 5G – CHALLENGES

Since obstacles must be overcome in order for new developments to succeed, 5G likewise has many difficulties to overcome. We can observe how swiftly radio technology advanced when we look at its evolution over time. Since 5G will be available in 2020 and 1G was first offered in the 1980s, the journey from 1G to 5G only took roughly 40 years. On the other hand, poor research methods, a lack of infrastructure and high costs have been the main difficulties we have faced along the road. The challenges brought on by 5G can be classified into two groups:

- Technical Challenges.
- Regular Obstacles.

a. **Technical Challenges**
   One of the most significant technological problems that need to be resolved is inter-cell interference. Due to their various sizes, conventional larger cells and concurrent small cells will interfere. Another is related to traffic control. Multiple Machine-to-Machine (M2M) devices in a cell could result in major system issues, such as issues with the radio access network (RAN), which would cause overload and congestion. Cellular networks, on the other hand, frequently only observe human-to-human activity (Mythili et al. 2017).

b. **Regular Obstacles**
   With Multiple Services: Providing services to a range of networks, technologies and devices that operate across multiple regions will be a major issue for 5G, in contrast to current radio signal services. All three of these functions – communication, navigation, and sensing – rely on the radio spec-

with 5G. Combining the two technologies will allow cloud applications to realize their full potential. Edge computing disrupts the tidy physical boundaries of the cloud data centre, forcing us to consider issues like security, scale, management, ownership and compliance. More critically, it exacerbates the scaling challenges with cloud-based management approaches.

REFERENCES

Blanco, B., Fajardo, J. O., Giannoulakis, I., Kafetzakis, E., Peng, S., Pérez-Romero, J., … & Xilouris, G. 2017. "Technology Pillars in the Architecture of Future 5G Mobile Networks: NFV, MEC and SDN." *Computer Standards & Interfaces* 54: 216–228.

Chettri, L., & Bera, R. 2019. "A Comprehensive Survey on Internet of Things (IoT) Toward 5G Wireless Systems". *IEEE Internet of Things Journal* 7(1): 6–32.

García, A. C., Maier, S., & Phillips, A. 2020. *Location-based services in cellular networks: From GSM to 5G NR*. Artech House.

Garg, N., Gupta, R., Kaur, M., Ahmed, S., & Shankar, H. 2023. "Efficient Detection and Classification of Orange Diseases using Hybrid CNN-SVM Model". *International Conference on Disruptive Technologies (ICDT)*: (721–726).

Gupta, R., Kaur, M., Garg, N., Shankar, H., & Ahmed, S. 2023. "Lemon Diseases Detection and Classification using Hybrid CNN-SVM Model". *Third International Conference on Secure Cyber Computing and Communication (ICSCCC)*: (326–331). IEEE.

Hooda, S., Lamba, V., & Kaur, A. 2021. "AI and Soft Computing Techniques for Securing Cloud and Edge Computing: A Systematic Review". *International Conference on Information Systems and Computer Networks (ISCON)*: (1–5).

Jyoti, Snehi, Manish, Snehi, & Rupali, Gill. 2011. "Virtualization as An Engine to Drive Cloud Computing Security." *International Conference on High Performance Architecture and Grid Computing*, Berlin, Heidelberg: (62–66).

Kaur, P., Garg, R., & Kukreja, V. 2023. "Energy-efficiency Schemes for Base Stations in 5G Heterogeneous Networks: A Systematic Literature Review". *Telecommunication Systems: Modelling, Analysis, Design and Management*, 84(1): 115–151.

Khuntia, M., Singh, D., & Sahoo, S. 2021. "Impact of Internet of Things (IoTs) on 5G". *Intelligent and Cloud Computing: Proceedings of ICICC 2019*. Springer Singapore. (125–136).

Loncar-Turukalo, T., Zdravevski, E., da Silva, J. M., Chouvarda, I., & Trajkovik, V. 2019. "Literature On Wearable Technology for Connected Health: Scoping Review of Research Trends, Advances, and Barriers." *Journal of Medical Internet Research* 21(9): e14017.

Mistry, I., Tanwar, S., Tyagi, S., & Kumar, N. 2020. "Blockchain for 5G-enabled IoT for Industrial Automation: A Systematic Review, Solutions, and Challenges." *Mechanical Systems and Signal Processing* 135: 106382.

Mythili, A., & Mahendran, S. 2017. "A Study of 5G Network: Structural Design, Challenges and Promising Technologies, Cloud Technologies". *International Journal of Advance Research, Ideas and Innovations in Technology* 3(6): 325–339.

Pons, M., Valenzuela, E., Rodríguez, B., Nolazco-Flores, J. A., & Del-Valle-Soto, C. 2023. "Utilization of 5G Technologies in IoT Applications: Current Limitations by Interference and Network Optimization Difficulties—A Review". *Sensors* 23(8): 3876.

Prashar, N., Hooda, S., & Kumar, R. 2023. "Current Status of Challenges in Data Security: A Review". In: Jain, R., Travieso, C.M., Kumar, S. (eds) *Cybersecurity and evolutionary data engineering*. ICCEDE 2022. *Lecture notes in electrical engineering*, (1073) Springer, Singapore. https://doi.org/10.1007/978-981-99-5080-5_1

Pujari, V., Patil, R., & Tambe, K. 2021. *Future of 5G wireless system*. Contemporary Research in India. ISSN (2231–2137).

Rodriguez, Jonathan. 2015. *Fundamentals of 5G mobile networks*. John Wiley & Sons.

Shetty, R. S. 2021. *5G mobile core network.* Apress.

Siddikov, I., Khujamatov, K., Reypnazarov, E., & Khasanov, D. 2021. "CRN and 5G based IoT: Applications, Challenges and Opportunities". *International Conference on Information Science and Communications Technologies (ICISCT)*: (1–5). IEEE.

Siriwardhana, Y., Porambage, P., Liyanage, M., & Ylianttila, M.. 2021. "A Survey on Mobile Augmented Reality With 5G Mobile Edge Computing: Architectures, Applications, and Technical Aspects". *IEEE Communications Surveys & Tutorials* 23(2): 1160–1192. Second quarter 2021. https://doi.org/10.1109/COMST.2021.3061981

Wübben, D., Rost, P., Bartelt, J., Lalam, M., Savin, V., Gorgoglione, M., ... & Fettweis, G. 2014. "Benefits and Impact of Cloud Computing on 5G Signal Processing". *IEEE Signal Processing Magazine* 31(6): 35–44.

Zhang, S.. 2019. "An Overview of Network Slicing for 5G". *IEEE Wireless Communications* 6(3): 111–117. https://doi.org/10.1109/MWC.2019.1800234

2 Architectural Design Principles for Smart and Sustainable Cities

Pratibha, Pummy Dhiman and Gaganpreet Kaur
Chitkara University Institute of Engineering & Technology,
Chitkara University, Rajpura, Punjab, India

2.1 INTRODUCTION

In the twenty-first century, smart and sustainable city planning is centred on architectural design concepts. The demand for creative and sustainable architectural solutions rises as cities grow to expand and face issues including population growth, resource depletion and environmental degradation (Jebaraj, Luke, et al., 2023).

Architecture design is essential to the development of smart cities since it shapes the urban environment to maximize sustainability, efficiency and resident quality of life. Through the integration of cutting-edge technologies and concepts of urban planning, architects are able to design environments that promote accessibility, resource management and connectivity (Su, Yiyi, Di Fan, 2023, 57).

To lessen its negative effects on the environment and improve the urban experience as a whole, smart design integrates features like intelligent transit systems, green infrastructure and energy-efficient buildings. Furthermore, careful planning can promote community involvement and social cohesiveness, which in turn can promote a feeling of shared duty and belonging in the city. Essentially, architecture design harmonizes technological innovation with human-centred urban development by acting as the foundation upon which the interconnected systems of a smart city are formed.

2.1.1 ARCHITECTURES OF A SMART CITY

Two forms of architecture may be found in a smart city: one that keeps focus on the city's outer features, such as its parks, avenues, and streets, and another that keeps a check on the inside of buildings, including the movement of people and commodities, the water and air conditioning systems, and other amenities. Figure 2.1 depicts the architecture of a smart city.

A smart city's internal architecture includes the control and command centres, sensor networks, AI and data analytics systems, and digital infrastructure that support the city's technical framework. This underlying design makes it possible to manage several urban systems centrally, gather and analyse data in real time, and communicate seamlessly. In sectors including transportation, energy management,

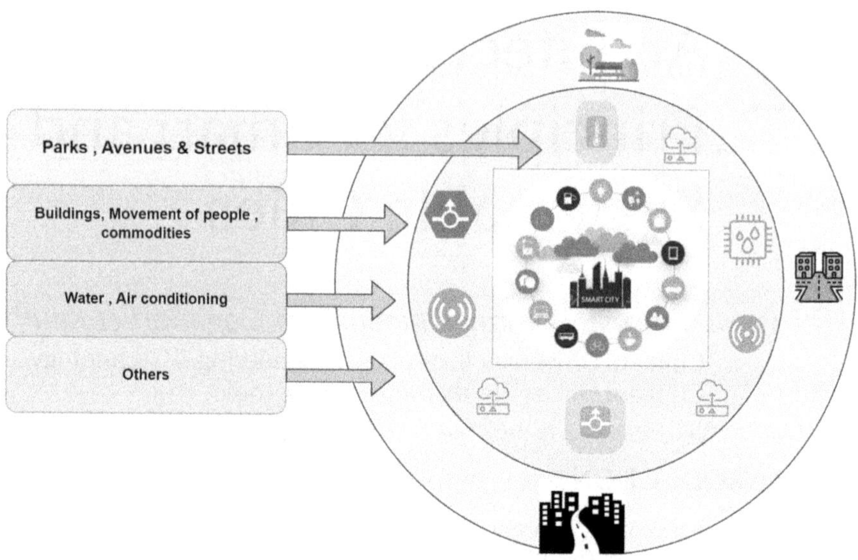

FIGURE 2.1   Architecture of smart city.

public safety and environmental sustainability, it forms the basis for streamlining urban operations, boosting overall efficiency and increasing service delivery.

The outer architecture of a smart city is the physical arrangement and infrastructure that shapes the urban landscape. This covers public spaces and services, transportation networks, energy-efficient buildings, clever land use planning, and urban planning. Creating dynamic, inclusive, and sustainable urban settings that enhance people's quality of life, promote mobility, and reduce their impact on the environment is the aim of exterior architecture. It covers the layout of roads, parks, buildings and infrastructure in addition to the inclusion of green spaces, features that promote foot traffic, and innovative transportation choices. All things considered, the physical structure of the city and the creation of a cohesive, living urban environment are formed by the external architecture.

The study is organized as follows (Figure 2.2): Section 2.2 discusses the smart city design principles following the introduction. Section 2.3 discusses the technologies and design advances that have influenced smart and sustainable city design. Section 2.4 discusses case studies of recent smart city projects, followed by a conclusion and future work in Section 2.5.

2.2 SMART URBAN DESIGN PRINCIPLES

Strategic planning techniques are part of smart urban design ideas, which are intended to build liveable, sustainable cities. In order to solve the issues that contemporary urban settings face, these principles place a strong emphasis on sustainability, connection, resilience, equity, innovation, mixed-use development and community participation. Below are some methods.

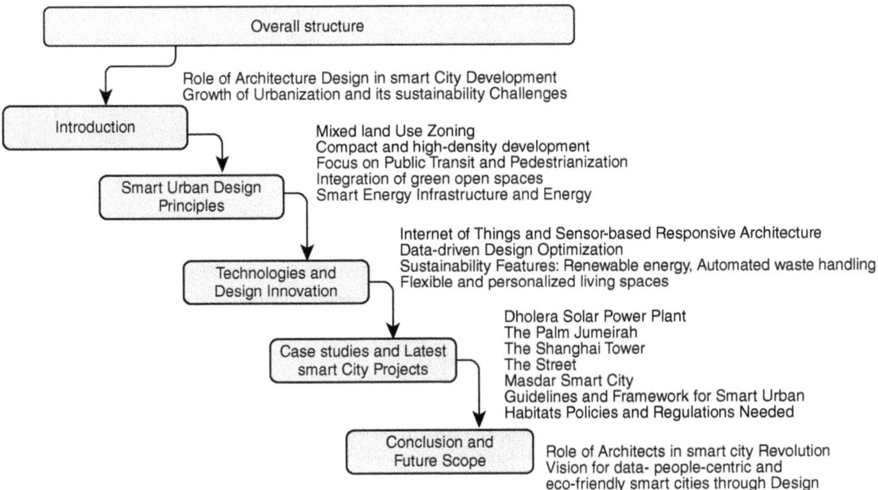

FIGURE 2.2   Overall structure of chapter.

2.2.1   MIXED LAND USE ZONING

It is a method of urban planning that encourages the coexistence of different land uses within a certain area or neighbourhood. Unlike discrete land use zoning (e.g., residential, business, and industrial), mixed land use zoning allows a combination of residential, commercial, recreational and occasionally industrial activities within the same community (Governing Smart Cities, 2021). The integration of residential, commercial and recreational spaces promotes walkability and accessibility and facilitates access to nearby job opportunities and essential services. This leads to a reduction in long commutes and dependence on cars. Furthermore, by establishing vibrant and active communities where people live, work, shop and interact nearby, it promotes interpersonal relationships, community cohesion and an enhanced quality of life. Furthermore, it optimizes land use efficiency by avoiding the allocation of large regions for single-use purposes, maximizing urban potential while reducing urban sprawl degradation (Jebaraj, Luke, et al., 2023).

Economically speaking, mixed zoning promotes small businesses, attracts capital and boosts activity. It reduces greenhouse gas emissions in terms of the environment by promoting alternative modes of transportation and protecting green spaces.

2.2.2   COMPACT AND HIGH-DENSITY DEVELOPMENT

"Compact and high-density development" is an urban planning approach that focuses on building densely populated areas with little space for sprawl. It involves building structures closer to one another and optimizing the amount of space that is available; vertical construction and mixed-use zoning are common examples of this. This tactic encourages the use of public transportation, reduces the need for lengthy excursions and improves walkability in the community (Abdelhamid, Bassant, et al., 2021).

Because compact and high-density development concentrates growth in a smaller area, it helps to preserve green spaces, maintain natural ecosystems and enhance environmental sustainability. Furthermore, it encourages vibrant and diverse communities, improves accessibility to utilities and services, and stimulates economic growth by optimizing the use of available land.

2.2.3 Focus on Public Transit and Pedestrianization

Developing effective public transportation networks and designing pedestrian-friendly urban spaces are top priorities when it comes to public transit and pedestrianization. This strategy seeks to lessen reliance on personal vehicles, lessen traffic and enhance air quality (Bakıcı et al., 2013).

Cities may give their citizens access to reasonably priced, dependable and easily accessible transportation choices by investing in public transportation infrastructure, such as light rail, buses and trains. Initiatives to pedestrianize cities also include altering areas to make walking and bicycling more important than driving. This entails dividing areas for pedestrians exclusively, extending sidewalks, adding bike lanes, and improving streetscapes by adding features like trees, benches and public art (Bakıcı et al., 2013). Cities may make their surroundings healthier, more sustainable and more liveable for their citizens while lowering carbon emissions and improving the general quality of life by encouraging public transportation and pedestrianization.

2.2.4 Integration of Green, Open Spaces

Including green, open spaces in urban settings is crucial to building liveable, sustainable and healthy cities. These areas include, among other things, waterfront promenades, parks, gardens, urban woods and green roofs (Jebaraj, Luke, et al., 2023). Cities may benefit their citizens and the environment in many ways by including green areas in their planning and design. Urban areas that include green, open spaces reduce the risk of flooding, manage stormwater runoff and offer shelter from natural disasters – all of which help to build more resilient societies. Wetlands, bioswales and green roofs may collect and store rainwater, relieving pressure on drainage systems and lowering the chance of flooding (Silva, Bhagya Nathali, et al., 2018, 38). Furthermore, green spaces act as natural barriers against extreme weather events like heat waves, storms and wildfires, enhancing a city's overall resilience.

2.2.5 Smart Energy Infrastructure and Buildings

Smart and advanced technologies are used in smart energy infrastructure and buildings to maximize energy efficiency, incorporate renewable energy sources, put energy storage systems into place and activate demand response programs. In addition, these systems integrate data analytics, smart grids, building automation and infrastructure for electric vehicles to lower carbon emissions, improve sustainability and foster resilience in energy management.

2.3 TECHNOLOGIES AND DESIGN INNOVATIONS

2.3.1 INTERNET OF THINGS AND SENSOR-BASED RESPONSIVE ARCHITECTURE

IoT sensors (Tekinerdogan, Bedir et al., 2023, 13) can be integrated into cities to collect data on waste management, energy consumption, traffic and air quality, enabling real-time modifications and decision-making for resource optimization and sustainability.

2.3.2 DATA-DRIVEN DESIGN OPTIMIZATION

Data-driven design optimization in smart and sustainable cities uses analytics (Dhiman, Kaur, 2022). Predictive modelling, parametric design, generative design, optimization algorithms, real-time feedback loops and visualization techniques to enhance urban design efficiency, environmental impact mitigation and quality of life (Kaur et al., 2022).

2.3.3 SUSTAINABILITY FEATURES – RENEWABLE ENERGY, AUTOMATED WASTE HANDLING

Urban design incorporates sustainability features like renewable energy integration and automated waste handling, promoting environmentally friendly practices (Tekinerdogan, Bedir et al., 2023, 13). Renewable energy reduces carbon emissions and fossil fuel reliance, while automated systems optimize resource utilization and minimize landfill waste.

2.3.4 FLEXIBLE AND PERSONALIZED LIVING SPACES

Living areas are planned to accommodate occupants' changing requirements and tastes while enhancing sustainability and resource efficiency. To make the most of available space, modular building methods, multipurpose furniture, smart house technologies and flexible interior treatments are employed. Wellness elements that support health and well-being include biophilic design ideas and programmable lighting. A thriving community is facilitated by shared amenities and common areas that encourage social interaction. These amenities provide residents with a sense of community and worth.

2.4 CASE STUDIES OF LATEST SMART CITY PROJECTS

This section explores some smart city initiatives, highlighting their transformative impact on infrastructure, sustainability and quality of life, highlighting their evolution as a beacon of urban innovation (Sharma, Pragya, Susheela Hooda, 2023). From regenerated urban cores to sustainable infrastructure solutions, these initiatives inspire communities to embrace the digital age while conserving their cultural identities and improving quality of life.

FIGURE 2.3   The Dholera solar power plant.

- Dholera Solar Power Plant
  In the Indian state of Gujarat sits the Dholera Solar Generating Plant, which ranks among the world's biggest solar generating facilities (Figure 2.3). Featuring a 700-MW capacity, it spans 11 square kilometres of land. Adani Green Energy constructed the facility, which optimizes solar energy collection by combining fixed and tracking panels. Rows of panels are linked to a central inverter system, which transforms the DC power the panels produce into AC power that can be sent into the grid.

  Many obstacles were encountered during the engineering of the Dholera Solar Power Plant. One of these was the extensive and uneven terrain that had to be traversed in order to install thousands of solar panels, inverters, transformers and other equipment. The project team overcame these obstacles by surveying the land and designing the plant layout using cutting-edge GIS technology (Sharma, Pragya 2024). To help transport supplies and heavy machinery, they also laid out a system of highways and railroads. They also used cutting-edge technology to reduce water use and increase energy output, making the facility extremely efficient and environmentally friendly.
- The Palm Jumeirah
  Palm Jumeirah is a massive building project that is part of a group of artificial islands called the Palm Islands, projected to be the largest in the world (Figure 2.4). Palm Jumeirah Island is famous for its exceptional architectural characteristics and groundbreaking engineering accomplishments, aiming to provide a luxurious living environment for residents and visitors alike. The construction of the man-made island was carefully planned to maximize the available coastline while also considering aesthetics. The complex network of roads and tunnels linking the island to the mainland plays a crucial role in fostering the city's growth and development (Sharma, Pragya, 2024). The engineers utilized millions of cubic meters of sand from the sea bed to construct the foundation of the island. The project showcases human intellect and invention. The project's distinctive design has greatly

FIGURE 2.4   The Palm Jumeirah.

boosted Dubai's economy and infrastructure, making it a profitable opportunity for the region.

- The Shanghai Tower
  The Shanghai Tower is a 632-meter-tall skyscraper in Shanghai, China. It is in the Pudong area. People know the tower for its unique shape, which was based on the Chinese symbols of a pagoda, a bamboo shoot, and a vase (Figure 2.5). The outside of the building is made up of curved glass panels that wrap around it and give it a lively, twisting shape. The tapered form of the tower also helps to lower wind loads, which keeps it stable when there is a lot of wind (Sharma, Pragya, 2024). The centre of the tower is made up of nine round steel tubes that are connected by outriggers and diagonal beams.

FIGURE 2.5   The Shanghai Tower architecture.

This system of supports is meant to give the tower the strength and security it needs to stand up to strong winds and earthquakes.

- The Street

The Street, designed by Sanjay Puri Architects, is tailored to the climate and location, offering diverse experiences and altering spatial perceptions across the 6-acre site. Implementing rainwater collecting, water recycling, solar panels, and natural ventilation make the project energy-efficient (Figure 2.6). All the buildings are positioned to create a view of expansive north-facing garden spaces that overlook a large playground (Goel, Payushi, 2024).

The hostel rooms also include wedge-shaped bay windows facing North. These elements aid in reducing heat absorption in accordance with the climate during the sun's presence in the Southern Hemisphere. During winter months in the Northern Hemisphere, direct sunlight protects rooms from being cold.

- Masdar Smart City

The ambitious United Arab Emirates-based project Masdar City aspires to build the first entirely sustainable metropolis in the world. It was proposed in 2007 as the first "zero emission" city, to be located less than 20 kilometres from Abu Dhabi. Various training facilities, research centres, alternative energy production centres, and specialist marketing and financing businesses will call the city home. The city is expected to be completely self-sufficient, with no pollution or garbage produced, and to house around 50,000 people (Sharma, Pragya, 2024).

A wholly owned subsidiary of Mubadala Development Company, Abu Dhabi Future Energy Company, is funding the project. Set in the United

FIGURE 2.6   The Street architecture.

Arab Emirates, the site of the world's biggest renewable energy conference, the idea is fresh and original. In order to improve system efficiency and generate new ideas, the Abu Dhabi Future Energy Company is actively seeking to construct a major future energy research centre. Intelligent energy, intelligent buildings, intelligent transportation, and intelligent people are the cornerstones of the endeavour. Smart buildings are designed to use renewable technologies on their rooftops to enable nearly zero energy use, and intelligent energy is provided by solar and wind power plants (Sharma, Pragya, 2 023). A vast network of micro-metropolitan to semi-individual transportation systems, such as the Rapid Transit System, enables convenient access around the city as part of smart mobility. Efficiency, renewable energy, and environmental sustainability are three areas where the city aspires to become a world leader in research and development.

2.4.1 GUIDELINES AND FRAMEWORK FOR SMART URBAN HABITATS

In order to build smart urban environments that prioritize efficiency, resilience and environmental protection while simultaneously improving the quality of life for inhabitants, it is necessary to combine technological advancements with sustainability and human-centred design principles. In order to create smart urban environments, consider the following:

- **United Nations' Sustainable Development Goals (SDGs)**: Coordinate smart urban habitat initiatives with SDGs (Sharma, 2024), with an emphasis on areas including responsible production and consumption, clean energy, climate action and sustainable cities and communities.
- **Resilience and Disaster Management**: When planning urban habitats, keep variables like climate change, natural catastrophes and infrastructural weaknesses in mind. In order to lessen the impact of potential dangers and make the city more resilient to unexpected events and stresses, it is important to have emergency response systems, adaptable infrastructure and backup plans (Goel, 2024).
- **Principles of Urban Planning and Design**: To maximize space utilization and minimize environmental impact, apply the principles of mixed land use, compact building design and efficient resource utilization. Promote walkability, alleviate traffic congestion and improve community well-being by highlighting pedestrian-friendly infrastructure, public transportation and green areas.
- **Utilizing Technology**: Automate urban services, make data-driven decisions, and enable real-time monitoring by integrating cutting-edge technologies like sensor networks, artificial intelligence (AI) and the Internet of Things (IoT) (Bonkra et al., 2021; Kiran et al., 2024). Put renewable energy systems, smart grids and energy-efficient technology into place to cut down on power usage and pollution.
- **Privacy and Data Governance**: Create strong frameworks for data governance to make sure that urban data is collected, stored and used ethically, all

while protecting the privacy and security of citizens. Increase participation from relevant parties in decision-making and promote openness and responsibility in data management (Berst et al., 2013).

- **Waste Management and the Circular Economy**: Push for waste reduction, resource optimization and sustainable consumption and manufacturing practices by using circular economy principles. To lessen the burden on landfills and protect our planet from contamination, try using creative waste management techniques like recycling, composting and upcycling (Saleh et al., 2024, 14).
- **Accessibility and Fairness**: Make sure that all residents have equal access to smart urban infrastructure, services and chances. Address digital inequalities and socioeconomic inequality by increasing access to affordable housing, education, healthcare, transportation and digital technologies (Gupta et al., 2024, 12).

2.4.2 POLICIES AND REGULATIONS NEEDED

Rules and regulations for smart cities are crucial for dealing with a range of problems that arise as cities expand. Regardless of their long-term goals, all smart cities should adhere to these principles because they represent essential conditions (Olaniyi et al., 2023, 25). To achieve the long-term goal of environmental sustainability, a city could, for instance, install smart lighting systems to cut down on energy consumption and pollution (Kaur et al., 2024, 2). But for the streetlights to remain on when needed, they must guarantee that the smart lighting is secure and resilient enough (Huda et al., 2024, 238). In order to establish a standard for good technology governance, the G20 Global Smart Cities Alliance drafted a set of policies that both politicians and tech companies may use as a guide. There are five guiding principles that form the basis of the guideline as shown in Figure 2.7.

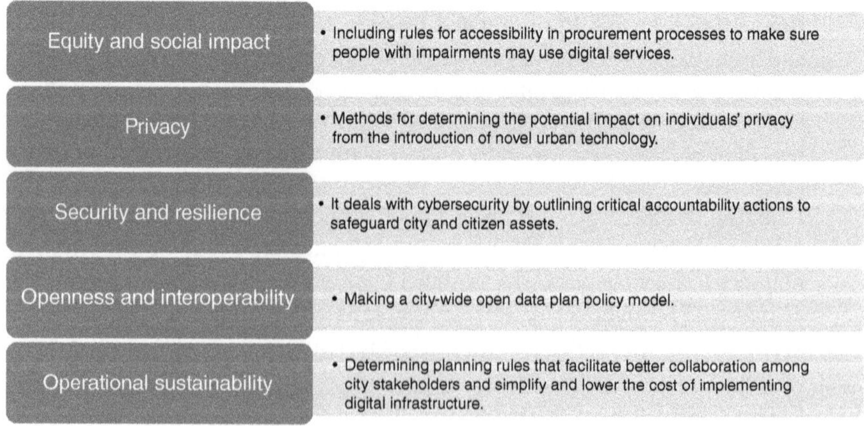

FIGURE 2.7   Guiding principles for smart city architecture.

2.5 CONCLUSION AND FUTURE SCOPE

Architectural design principles are crucial in creating smart and sustainable cities. This book chapter has examined the essential concepts of architectural design in smart and sustainable cities, examining how to create technologically sophisticated and ecologically conscious urban landscapes. Architects are visionary leaders who blend innovation with design to create urban spaces that cater to modern society's evolving needs. They should focus on data-driven, people-centric and eco-friendly design, utilizing technology and analytics to optimize urban planning and improve quality of life. People-centric design prioritizes accessibility, inclusivity and social cohesion, fostering community engagement and wellness. Environmental stewardship and ecological preservation are central to this vision. The future of architectural design for smart and sustainable cities holds immense promise. As technology evolves, architects will be called upon to adapt and innovate. Advanced technologies like artificial intelligence, IoT and augmented reality can revolutionize design processes. Regenerative design, incorporating principles of circularity, biomimicry and adaptive reuse, can contribute to environmental sustainability. Architects will also play a significant role in advocating for social equity and inclusion. By fostering cross-disciplinary collaboration, architects can continue to shape urban landscapes in innovative and transformative ways.

REFERENCES

Abdelhamid, Bassant, Awab Al-Habal, and Ahmed Khattab. 2021. "Energy-efficient Layered IoT Smart Home System." In Ibrahim El Dimeery, Moustafa Baraka, Syed M. Ahmed, Amin Akhnoukh, Mona B. Anwar, Mahmoud El Khafif, Nagy Hanna, Amr T. Abdel Hamid (eds.) *Design and Construction of Smart Cities: Toward Sustainable Community*, Springer International Publishing: pp. 170–177.

Bakıcı, Tuba, Esteve Almirall, and Jonathan Wareham. 2013. "A Smart City Initiative: The Case of Barcelona." *Journal of the Knowledge Economy* 4: 135–148.

Berst, J., L. Enbysk, C. Caine, C. Williams, and T. Davis 2013. *Smart Cities Readiness Guide: The Planning Manual for Building Tomorrow's Cities Today*. Smart City Council. Available from: http://smartcitiescouncil.com/resources/smart-cities-readiness-guide (Accessed: February 10, 2024).

Bonkra, Anupam, and Pummy Dhiman. 2021. "IoT Security Challenges in Cloud Environment." In *2021 2nd International Conference on Computational Methods in Science & Technology (ICCMST)*: pp. 30–34. IEEE.

Dhiman, Pummy, and Amandeep Kaur. 2022. "Significance of Convolutional Neural Network in Fake Content Detection: A Systematic Survey." In Jain, S., Marriwala, N., Tripathi, C.C., Kumar, D. (eds.) *Emergent Converging Technologies and Biomedical Systems*. ETBS 2022. Lecture Notes in Electrical Engineering, vol 1040. Springer, Singapore. https://doi.org/10.1007/978-981-99-2271-0_26

Goel, Payushi. 2024. "10 Indian Architects Designing with Innovative Technology - RTF: Rethinking the Future, RTF | Rethinking the Future." *RTF | Rethinking the Future*. Available at: https://www.re-thinkingthefuture.com/rtf-fresh-perspectives/a1880-10-indian-architects-designing-with-innovative-technology-2/ (Accessed: February 20, 2024).

Gupta, Manisha, and Himani Gupta. 2024. "Sustainable Urban Development of Smart Cities in India-a Systematic Literature Review." In Alejandra Manco Vega, Gala García Reátegui, Iemima Ploscariu, Kamel Belhamel, Leena Shah, Paula Marjamäki, Sonja Brage (eds.) *Sustainability, Agri, Food and Environmental Research*. Universidad Católica de Temuco: p. 12.

Huda, Noor Ul, Ijaz Ahmed, Muhammad Adnan, Mansoor Ali, and Faisal Naeem. 2024. "Experts and Intelligent Systems for Smart Homes' Transformation to Sustainable Smart Cities: A Comprehensive Review." *Expert Systems with Applications*. 238: 122380.

Jebaraj, Luke, Alex Khang, Vadivelraju Chandrasekar, Antony Richard Pravin, and Kumar Sriram. 2023. "Smart City: Concepts, Models, Technologies and Applications." In Khang, A., Gupta, S.K., Rani, S., & Karras, D.A. (eds.) *Smart Cities*: pp. 1–20. CRC Press.

Kaur, Amandeep, Meenu Khurana, and Robertas Damaševičius. 2024. "Multimodal Hinglish Tweet Dataset for Deep Pragmatic Analysis." *Data* 9(2): 38.

Kaur, Gaganpreet, Amandeep Kaur, and Meenu Khurana. 2022. "A Review of Opinion Mining Techniques." *ECS Transactions* 107(1): 10125.

Kiran, Vidhu, Susheela Hooda, and Tejinder Kaur. 2024. "Various Possible Attacks on Internet of Things (IoT)." In Arvind Dagur, Karan Singh, Pawan Singh Mehra, D. K. Shukla (eds.) *Artificial Intelligence, Blockchain, Computing and Security. Volume 2*: pp. 141–146. CRC Press.

Olaniyi, Oluwaseun, Olalekan J. Okunleye, and Samuel Oladiipo Olabanji. 2023. "Advancing Data-driven Decision-making in Smart Cities Through Big Data Analytics: A Comprehensive Review of Existing Literature." *Current Journal of Applied Science and Technology* 40(25): 10–18.

Saleh, F., A. Elhendawi, A. S. Darwish, P. Farrell 2024. "An ICT-based Framework for Innovative Integration between BIM and Lean Practices Obtaining Smart Sustainable Cities." *International Journal of BIM and Engineering Science (IJBES)* 14: 68.

Sharma, Bhavna, and Susheela Hooda. 2023. "AI-Based Prediction Models for Network Security Attacks: A Comparative Study." Available at SSRN 4638515.

Sharma, Pragya. 2024. "6 most Innovative Architecture Mega Projects Around the World." Novatr Prev OX. Available at: https://www.novatr.com/blog/impressive-architecture-mega-projects (Accessed: February 26, 2024).

Silva, Bhagya Nathali, Murad Khan, and Kijun Han. 2018. "Towards Sustainable Smart Cities: A Review of Trends, Architectures, Components, and Open Challenges in Smart Cities." *Sustainable Cities and Society* 38: 697–713.

Su, Yiyi, and Di Fan. 2023. "Smart Cities and Sustainable Development." *Regional Studies* 57(4): 722–738.

Tekinerdogan, Bedir, Ömer Köksal, and Turgay Çelik. 2023. "System Architecture Design of IoT-Based Smart Cities." *Applied Sciences* 13(7): 4173.

World Economic Forum. 2021. "Governing Smart Cities: Policy Benchmarks for Ethical and Responsible Smart City Development." *World Economic Forum*. https://www.weforum.org/publications/governing-smart-cities-policy-benchmarks-for-ethical-and-responsible-smart-city-development/

3 Various Frameworks for Smart City and Urbanization System

Poonam, Kamal Kumar, Varsha Rani and Rupinder Kaur
Ch. Devi Lal State Institute of Engineering & Technology, Sirsa, India

3.1 INTRODUCTION

Smart cities and urbanization are complex processes that require careful planning and implementation to be successful. There are many different methodologies and frameworks that can be utilized when developing smart city initiatives and managing urban growth as shown in Figure 3.1.

The Netherlands' Amsterdam has adopted smart city applications that centre on public involvement, energy-saving smart lighting controls, enhanced safety, reduced traffic, digital platforms for public interaction and open databases. These correspond with the indicators of clever people (SP), smart mobility (SM), smart environment (SEnv) and smart lifestyle (SL). Improving life quality and involving citizens are important objectives. Similarly, the city of Groening in the Netherlands is pursuing energy-efficient systems and real-time transportation data analytics as core to its smart city approach. The goals connect to the sustainable environment (SEnv) and smart mobility (SM) indicators. Energy usage optimization and convenient transportation are priorities.

Nice, France introduces smart applications for lighting, road circulation optimization, waste management and environmental monitoring. These link to sustainable environment (SEnv), smart mobility (SM) and smart living (SL) indicators. The objectives are to improve sustainability and standard of living through technology.

Padova, Italy prioritizes environmental monitoring via smart sensors as its primary smart city focus. This aligns solely with the sustainable environment (SEnv) indicator. Comprehensive real-time data aims to guide decisions for natural resource conservation.

Turin, Italy has created a digital portal with services and information for citizen engagement, focused on user-centric design and accessibility for public services. It has also rolled out smart meters across the city. These efforts tie into the sustainable growth (SG), sustainable environment (SEnv) and smart people (SP) goals. Empowering people is the impetus.

DOI: 10.1201/9781003467892-3

Smart Economy (Competitiveness)	Smart Governance (Participation)	Smart People (Social and Human Capital)
• Innovative spirit • Entrepreneurship • Economic Image & trademarks • Productivity • Flexibility of labour market • International embeddedness • Ability to transform	• Participation in decision-making • Public and social services • Transparent governance • Political strategies & perspectives	• Level of qualification • Affinity to life-long learning life to Social and ethnic plurality • Flexibility • Creativity • Cosmopolitanism/Open-mindedness • Participation in public life
Smart Mobility (Transport and ICT)	Smart Mobility (Natural resources)	Smart Quality (Quality of life)
• Local accessibility • (Inter-)national accessibility • Availability of ICT-Infrastructure • Sustainable, Innovative and safe transport systems	• Attractivity of natural conditions • Pollution • Environmental protection • Sustainable resource management	• Cultural facilities • Health conditions • Individual safety • Housing quality • Education facilities • Touristic attractivity • Social cohesion

FIGURE 3.1 Smart city indicators (SCI) (based on National Institute of Urban Affairs).

One of the world's most extensive smart city approaches is found in Barcelona, Spain. It incorporates open data platforms, water conservation, smart public transportation, green technologies, participatory governance and an ecosystem of entrepreneurs. This includes metrics for smart people (SP), smart living (SL), smart mobility (SM), smart economy (SEcon) and sustainable growth (SG). The variety of programs aims to improve living quality and sustainability. Bilbao, Spain focuses its smart city efforts on citizen participation, smart parking systems, open data platforms and broader open data access. These link to sustainable growth (SG) for engagement, smart mobility (SM) for parking and smart people (SP) for transparency. Grassroots citizen input complements technology innovations.

The Spanish city of Santander Parallel's Bilbao's model with citizen participation processes, smart parking, and digital literacy platforms checking the boxes for sustainable growth (SG), smart mobility (SM) and smart people (SP) indicators. Removing barriers to technology usage is embedded in the approach.

Stockholm, Sweden prioritizes connectivity and economic growth in its smart city strategy. It has invested in extensive fibre optic and wireless networks to enable business expansion. This aligns with the smart economy (SEcon) and smart living (SL) indicators. State-of-the-art connectivity promotes innovation and convenient digital services.

Malmo, Sweden focuses on citizen centricity in its smart city approach. It has created digital participation platforms and open data portals where citizens actively provide input on decision-making. These tie into the sustainable growth (SG) and smart living (SL) strategic goals.

Copenhagen, Denmark takes a comprehensive approach spanning citizen engagement, broadband and mobile service growth, smart energy, upgraded transport systems, and open data. This hits the sustainable growth (SG), sustainable environment (SEnv), smart mobility (SM), smart economy (SEcon), smart people (SP) and smart living (SL) indicator targets. The goals are ambitious – to become carbon neutral by 2025.

Brno in the Czech Republic concentrates its smart city goals on promoting entrepreneurship and innovation. This connects solely to the smart economy (SEcon) indicator. Facilitating business growth is viewed as a catalyst for jobs and economic development.

Plan IT Valley in Portugal has commenced the introduction of sensor networks for real-time monitoring of environmental conditions. This links to the sustainable environment (SEnv) strategic indicator. The focus is on sustainability through data-driven insights.

Similarly, Porto, Portugal created an innovation hub to spur technology entrepreneurship, tying into the smart economy (SEcon) indicator for business expansion.

Norfolk in the United Kingdom aligns with Plan IT Valley in Portugal – leveraging data applications for environmental monitoring, while also boosting citizen engagement. These link to the sustainable growth (SG) and sustainable environment (SEnv) strategic goals.

The scope of smart city projects in London, England is similar to that of Barcelona, Spain. These initiatives include assistance for entrepreneurship, digital platforms, open data programs, pollution and congestion reduction, public participation, and the deployment of green technology in parking and transportation. Sustainable growth (SG), sustainable environment (SEnv), smart mobility (SM), smart economy (SEcon), smart people (SP) and smart living (SL) are the six indicators that these relate to. In Belgium, Liege has initiated an innovation lab called Venture Lab to incubate startups, tying into the smart economy (SEcon) strategic indicator for business productivity.

Vienna, Austria focuses smart city efforts primarily on energy efficiency improvements leveraging smart sensors and coordination of mobility systems. This links to the sustainable environment (SEnv) and smart mobility (SM) outcome goals.

Friedrichshafen in Germany takes a citizen-centric approach with applications spanning distress signal notifications, health monitoring, metering infrastructure, upgraded transport and open data platforms. The applications align with goals across five indicators – sustainable environment (SEnv), smart mobility (SM), smart economy (SEcon), smart people (SP) and smart living (SL).

Busan in South Korea heavily focuses on shared digital infrastructure and data capabilities with initiatives ranging from shared apps, geographic information systems, mobility analytics, open data platforms, and more. These support the smart mobility (SM), smart economy (SEcon), smart people (SP) and smart living (SL) outcome objectives.

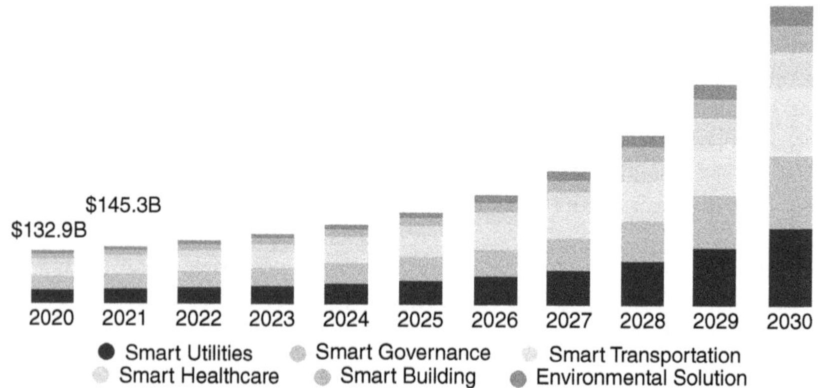

FIGURE 3.2   The US smart cities market size.

(Source: Zhao and Zhang, 2019.)

Fellow South Korean city Songdo applies technology across the board to sustainability and quality of life improvements, including energy efficiency, digital connectivity, automation, waste and lighting systems management. The goals span sustainable environment (SEnv), smart mobility (SM), smart economy (SEcon), smart people (SP) and smart living (SL) indicators.

Pune, India launched open data, environmental sensors and public WiFi as initial step toward becoming a smart sustainable city. These tie into the sustainable growth (SG), smart economy (SEcon), smart mobility (SM), smart people (SP) and smart living (SL) outcome categories.

Dholera, India is taking a sustainable economic growth approach with its smart city strategy focused on integrated transit and logistics hubs. This links to the smart mobility (SM) and smart economy (SEcon) impact indicators. The US Smart Cities market size has been shown in Figure 3.2.

Fujisawa, Japan aligns with Pune, India in its concentration on emissions reduction and mobility management to achieve environmental sustainability and transport optimization. These efforts connect to sustainable environment (SEnv) and smart mobility (SM) goals.

Shanghai, China similarly focuses first on sustainability through crowd flow control, infrastructure improvements and digital platforms to manage urban density. This matches to smart mobility (SM), smart people (SP) and smart living (SL) outcome objectives.

3.2 TOP-DOWN VERSUS BOTTOM-UP APPROACHES

One of the first decisions that needs to be made when embarking on a smart city or urban planning project is whether to take a top-down or bottom-up approach.

Top-down methodologies are centralized, government-led initiatives. City leaders create a vision and strategy for the smart city, which is then executed in a systematic fashion. These types of projects can allow for large-scale, rapid deployment of infrastructure and technology. However, they have been criticized for not always adequately incorporating citizen and stakeholder input.

Bottom-up approaches are more grassroots, citizen-led movements. Local innovators, entrepreneurs, and community groups organize and implement smart city solutions. This can create more organic, hyperlocal innovations that reflect the specific needs of citizens. However, bottom-up efforts can lack coordination and can encounter challenges scaling citywide.

Many experts argue that a combination of top-down and bottom-up methodologies is required for optimal smart city development. Central leadership can enable integration and standardization, while community involvement allows for contextualization and inclusive innovation.

3.3 TECHNOLOGY-FIRST VERSUS HUMAN-CENTRED DESIGN

Another consideration is whether to pursue a technology-first or human-centred approach to smart city design.

With a technology-first methodology, cities focus first on deploying emerging technologies such as IoT sensors, big data analytics and autonomous vehicles. The assumption is that these technologies will inherently improve efficiency, sustainability and quality of life. However, critics point out that applying technology without considering human contexts can fail to solve real problems or even create new issues.

Human-centred design puts the experiences of citizens at the forefront. Urban planners research people's needs and design solutions to fit those needs, utilizing technology as appropriate. While this approach requires more upfront research, it is argued to create more holistic, citizen-oriented outcomes.

3.4 INTEGRATED VERSUS SPECIALIZED STRATEGIES

Smart city initiatives can either utilize an integrated strategy that looks at the city as an interconnected system, or a more specialized strategy that focuses on specific sectors like transportation, energy or housing.

With an integrated approach, city systems are designed to share data and operate in sync with one another. For example, transportation and power systems may communicate real-time data to optimize traffic flow and energy usage citywide. This requires extensive coordination but can uncover synergies and efficiencies between sectors.

In a specialized strategy, planning happens within individual city domains. Transportation planners may implement intelligent traffic management systems while housing officials build affordable smart apartments. This approach allows for targeted problem-solving but can miss cross-cutting solutions.

Most experts recommend a combination of both strategies. An overarching integrated vision can enable big picture planning, while specialized approaches allow for in-depth innovation within sectors.

3.5　OPEN INNOVATION VERSUS CLOSED SYSTEMS

Smart city developers also need to determine whether to pursue open or closed technological systems.

With open innovation, cities release data and APIs openly and encourage collaboration with external third-party developers to create urban apps and services. This spurs innovation and experimentation but can lack security and quality control.

In closed systems, city governments maintain control over technology standards and development. This allows for greater security and oversight but inhibits external innovation.

Many leading smart cities aim for a middle ground, using common data standards and architecture while enabling external innovation through open APIs and hackathons. This balances openness and consistency.

3.6　AGILE VERSUS LINEAR DEVELOPMENT

The software development methodology used to build smart city technologies is another key consideration. A linear, sequential approach moves through defined stages like planning, design, testing and deployment. Agile methodologies take an iterative approach, continuously integrating user feedback and rapidly evolving products.

Linear methods enable large-scale consistency and standardization. However, they can struggle to adapt to changing citizen needs and new technologies over a city's multi-decade evolution.

Agile provides constant refinement and citizen-orientated features. But it can be harder to coordinate across city agencies and partners. Experts suggest combining elements of both approaches.

3.7　PUBLIC–PRIVATE PARTNERSHIPS

Public–private partnerships (PPPs) are collaborations between city governments and private sector companies to deliver smart city projects. PPPs allow cities to tap into the expertise and technology of innovative companies. The companies in turn gain new business opportunities and market access.

PPPs can accelerate smart city development and enable cities to take on complex, capital-intensive projects. But cities also need to ensure proper oversight of vendors to guarantee solutions truly serve the public interest. Examples of smart city PPPs include Cisco's partnerships with cities like Paris and Adelaide.

3.8　LIVING LAB METHODOLOGY

The living lab model, as shown in Figure 3.3, is a user-centric innovation methodology increasingly used in smart city design. Living labs utilize public/private partnerships to prototype and validate smart city technologies and services in real urban environments with actual citizen engagement and feedback.

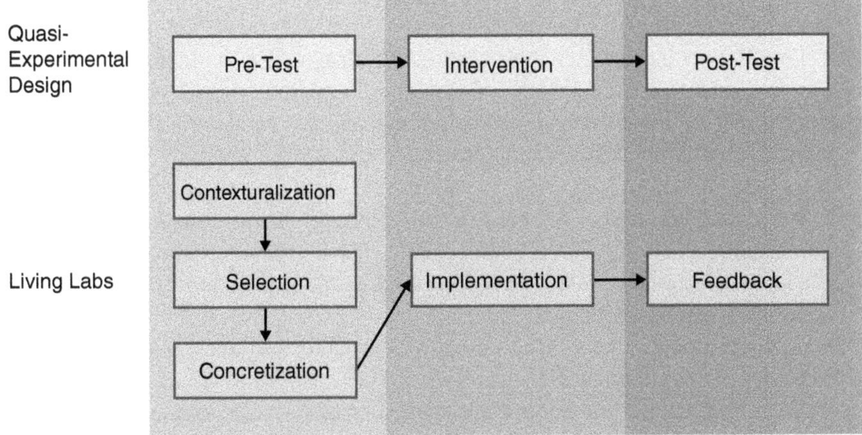

FIGURE 3.3   Living lab flow model.

(Source: Schuurman et al., 2016.)

For example, living labs may deploy connected IoT infrastructure in a neighbour-hood and gather resident feedback to refine technologies and experiences. Living labs allow new smart city solutions to be co-created with citizens and stakeholders rather than imposed top-down. Barcelona's living lab strategy is a pioneering example.

3.9   SMART CITY ECOSYSTEM APPROACH

A smart city ecosystem perspective recognizes that a city comprises complex economic, social and technical systems that interact and evolve together. By taking a big picture systems-level approach, planners can design smarter initiatives that enhance the overall urban ecosystem.

This methodology utilizes tools like system dynamics modelling, agent-based modelling and social network analysis to understand relationships and interactions between city subsystems. It provides a more dynamic, adaptive view compared to traditional linear planning. Singapore is a leader in system-wide smart city design.

3.10   GEOSPATIAL INTELLIGENCE FRAMEWORK

Geospatial intelligence (GEOINT) uses location-based data from GIS mapping, satellites, drones and GPS to provide insights into patterns and trends across a city. GEOINT enables data visualization and advanced analytics to aid planning and operations.

For example, geospatial data analytics can optimize transit networks based on real-time traffic flows. Or health officials can track disease outbreaks and air pollution levels using real-time maps. A GEOINT framework integrates 3D maps with IoT sensors, computer vision and big data analytics.

3.11   INTEROPERABILITY STANDARDS

Interoperability is a critical enabler of effective smart cities. It allows diverse urban subsystems and technologies to connect and share data seamlessly. To achieve interoperability, cities must establish common standards and architectures.

Some key interoperability standards used in smart cities include:

- IoT standards like oneM2M allow physical devices and sensors to interconnect.
- Data standards like FIWARE NGSI provide uniform data formats and APIs for information exchange.
- Open protocols like MQTT and CoAP enable resource-constrained IoT devices to communicate efficiently.
- Security standards like mutual authentication protect data exchange between systems.

Interoperability standards enable urban systems to exchange real-time data, allowing for intelligent integrated operations and advanced analytics.

3.12   SMART CITY MATURITY MODEL

The smart city maturity model is a framework for assessing a city's progress through successive stages of smart city development. Developed by analysts like Frost & Sullivan, the model defines key dimensions like:

- Smart mobility – Intelligent transport systems.
- Smart utilities – Energy, water, waste management.
- Smart governance – Digital services, citizen engagement.
- Smart infrastructure – Connected public assets and operations.

The model provides an evaluation and planning tool to guide cities through rising levels of maturity – from ad-hoc pilots to fully optimized and interconnected systems. It can benchmark cities globally while setting milestones for advancement.

3.13   FLAWED AND LIMITING METHODOLOGIES

While the methodologies discussed so far represent best practices, some flawed or limiting approaches should also be avoided:

- Technocratic planning that ignores public input and social/behavioural factors.
- Over-Reliance on closed proprietary systems that stifle innovation.
- Lack of interoperability and data standards that inhibit integration.
- Prioritizing cost savings over actual citizen benefits and social value.
- Focusing too much on optimizing the present rather than building long-term resiliency and adaptability.

Smart city development requires holistic thinking and avoiding narrow technocratic solutions. The most successful smart cities utilize a blend of strategies tailored to local contexts.

3.14 CASE STUDIES

Singapore: The Model of Top-Down, Integrated Systems Planning. Singapore is renowned as one of the most advanced smart cities globally. This has been achieved through systematic top-down planning by the government focused on integrating complex urban systems.

The Smart Nation and Digital Government Office spearheads smart city efforts in coordination with agencies covering the economy, health, transport, housing and sustainability. Rather than taking a piecemeal approach, Singapore has laid out comprehensive strategies for how smart technologies can improve livelihoods.

For example, the Smart Nation Sensor Platform deploys sensors and IoT infrastructure across the island to collect real-time data and enable automated smart services. An emphasis is placed on integrating and making sense of data holistically using analytics and modelling.

Citizens are engaged through channels like the Smart Nation Fellowship and the use of behavioural insights to develop citizen-centric digital services. The Impact of Smart City Planning and Construction on Economic and Social Benefits Based on Big Data Analysis is shown in Figure 3.4. While still government-driven, Singapore's meticulous approach has created a vibrant smart city ecosystem.

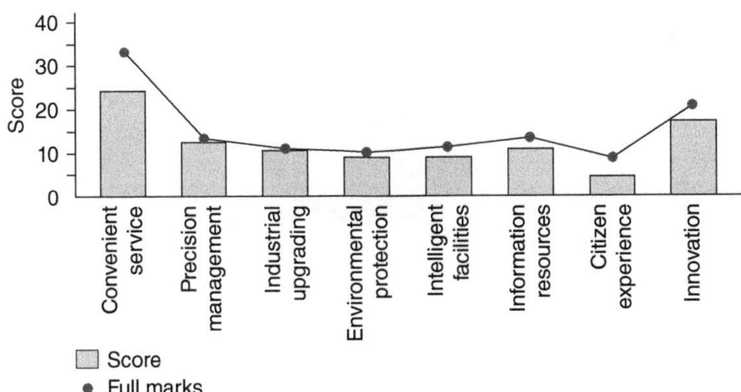

FIGURE 3.4   Impact of smart city planning and construction on economic and social benefits based on big data analysis.

(Source: Grand View Research, 2022.)

3.15 BARCELONA: GRASSROOTS INNOVATION THROUGH LIVING LABS

Barcelona exemplifies a human-centred, bottom-up approach to smart cities. The city pioneered the use of living labs – localized zones to test out smart technologies with real citizens before scaling up citywide. The Living Labs has been shown in Figure 3.5.

For example, Barcelona's living labs have tested out:

- Smart streetlights that adapt to ambient conditions and use presence sensors.
- Environmental sensors that measure noise, traffic and pollution in granular detail.
- Smart parking technology linking users and real-time spot availability.
- Digital kiosks providing hyperlocal information to residents.

These pilots have evolved into long-term platforms mobilizing citizens, entrepreneurs and researchers to keep generating ideas and technologies that enhance lives. Barcelona smart city projects focus first on citizen needs rather than technology, leading to more inclusive innovation.

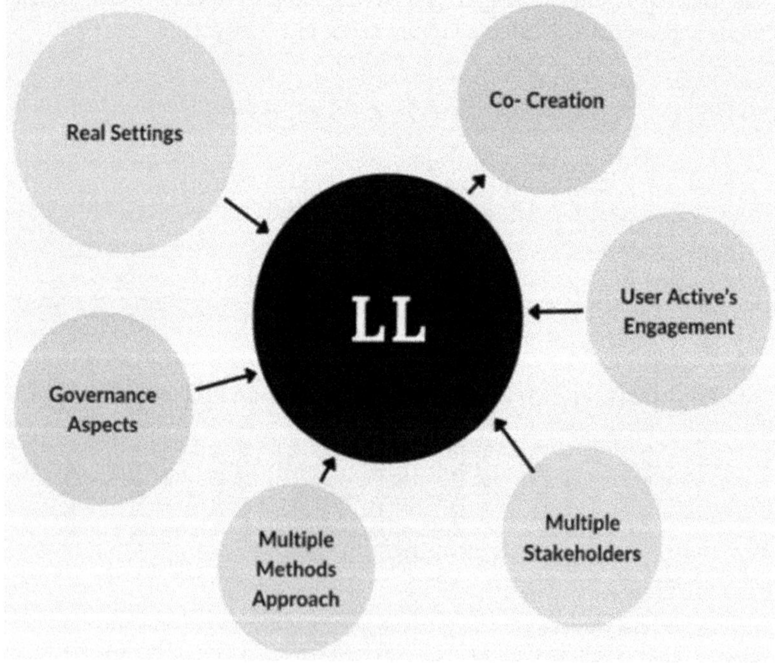

FIGURE 3.5   Living labs: Science, society and co-creation.

(Source: Giffinger et al., 2007.)

3.16 CHICAGO: PIONEERING CITY DATA SHARING

Chicago has led the way in using open data to drive smart city development. In 2010, Chicago became the first city to set up an open data portal with machine-readable datasets available to all. This spawned the creation of over 750 civic apps and services by external developers.

For example, apps were built to find nearby transit options, access business permits, track lobbyist activity, and crowd source neighbourhood issues. These digital services have improved citizen access to city resources and engagement.

Chicago also launched the Array of Things project, an urban sensing network that collects real-time environmental and activity data across the city. This massive data asset is shared openly while preserving privacy. Chicago's data-driven approach drives accountability, economic development and urban problem-solving.

3.17 CONCLUSION

In the realm of smart city initiatives, a rich tapestry of frameworks and methodologies exists, and their selection largely hinges on the unique local contexts and specific objectives of the project at hand. These methodologies span a wide spectrum, each offering its own set of advantages and applications.

One overarching approach that cities often find valuable is the integration of cross-cutting methodologies. This approach is akin to adopting a holistic systems perspective, allowing cities to view their urban development as an interconnected web of various facets. By doing so, cities can unlock the potential for synergy between different sectors, fostering a more efficient and effective smart city ecosystem. In essence, it's about recognizing that the various components of a city – transportation, energy, healthcare, education and more – are intricately linked, and optimizing one can have ripple effects across the entire urban landscape.

For more targeted efforts, specialized strategies come into play. These methodologies zero in on specific domains or areas within the smart city framework. They enable cities to craft precise plans and solutions tailored to the unique challenges and opportunities presented by these focal points. For instance, a city might employ a specialized strategy for energy efficiency, transportation optimization or public safety enhancement. This level of granularity allows cities to address critical issues more directly and comprehensively.

However, it's not just about top-down planning and domain-specific strategies. Smart cities today often embrace hybrid models. These models blend top-down governance, typically led by city authorities and policymakers, with bottom-up grassroots innovation driven by the citizens themselves. This approach ensures that the solutions and initiatives implemented in a smart city are inclusive and citizen-centric. After all, the ultimate goal of a smart city is to enhance the quality of life for its residents, and involving them in the decision-making process is vital for success.

In the quest for smart urbanization, another crucial aspect is investing in interoperability and open standards. These principles underpin the scalability of smart city solutions. Without interoperability – the ability of different systems and devices to communicate and work together seamlessly – the potential of a smart city is severely limited. Open standards ensure that technologies and solutions are not locked into proprietary silos, fostering innovation and competition while making it easier to integrate new technologies as they emerge.

In conclusion, smart city planners face a dynamic landscape of methodologies, each with its strengths and applicability. The key to success lies in carefully assessing the local context, project goals, and the specific challenges at hand. By doing so, cities can develop sustainable, intelligent urban environments that cater to the unique needs of their residents, while also ensuring that they remain adaptable and forward-looking in the face of evolving technologies and urban dynamics.

BIBLIOGRAPHY

Alisjahbana, A.S. 2019. Smart Cities Hold the Key to Sustainable Development [Online]. https://www.unescap.org/op-ed/smart-cities-hold-key-sustainable-development

Bhattachary, S., and Rathi, S. 2015. *Reconceptualising Smart Cities: A Reference Framework for India*. CSTEP-Report-2015-03.

Brynjolfson, E., McAfee, A., and Spense, M. 2019. "New World Order: Labor, Capital and Ideas in the Power Law Economy". Foreign Affairs [Online]. https://www.foreignaffairs.com/articles/united-states/2014-06-04/new-world-order

Ellsmoor, J. 2019. Smart Cities: The Future of Urban Development. Forbes [Online]. https://www.forbes.com/sites/jamesellsmoor/2019/05/19/smart-cities-the-future-of-urban-development/#327e8ac32f90

Executive Office of the President. 2016. Report to the President, Technology and the Future of Cities. President's Council of Advisors on Science and Technology [Online]. https://www.whitehouse.gov/sites/whitehouse.gov/files/images/Blog/PCAST%20Cities%20Report%20_%20FINAL.pdf

Giffinger, R., Fertner, C., Kramar, H., Kalasek, R., Pichler-Milanovic, N., and Meijers, E. J. 2007. Smart cities. Ranking of European medium-sized cities. Final Report.

Grand View Research. 2022. "Smart cities market size, share & trends analysis report by application (smart utilities, smart governance, smart transportation, smart healthcare), and segment forecasts, 2024 – 2030." https://www.grandviewresearch.com/industry-analysis/smart-cities-market

Green, B. 2019. *"The Smart Enough City: Putting Technology in Its Place to Reclaim Our Urban Future."* MIT Press.

Hooda, S., Lamba, V., and Kaur, A. 2021. "AI and Soft Computing Techniques for Securing Cloud and Edge Computing: A Systematic Review". *5th International Conference on Information Systems and Computer Networks (ISCON)*, Mathura, India. (pp. 1–5). https://doi.org/10.1109/ISCON52037.2021.9702422

Hooda, Susheela. 2021. "A New Approach for Power-Aware Routing for Mobile Adhoc Networks Using Cluster Head with Gateway Table." *International Journal of Web-Based Learning and Teaching Technologies (IJWLTT)*, 16(4). pp. 47–59. https://doi.org/10.4018/IJWLTT.20210701.oa4

International Telecommunication Union. 2019. ICTs for a Sustainable World #ICT4SDG[Online]. https://www.itu.int/en/sustainable-world/Pages/default.aspx

Johannes, M. 2019. Smart City and Urbanization Challenges [Online]. https://www.e-zigurat.com/blog/en/smart-cities-urbanization-challenges/

Kang, Lei and Siyou, Xia. 2023. "Study on Urbanization Sustainability of Xinjiang in China: Connotation, Indicators and Measurement". *International Journal of Environmental Research and Public Health* 20(3). https://doi.org/10.3390/ijerph20032535

Kumar, A., Hooda, S., Gill, R., Ahlawat, D., Srivastva, D. and Kumar, R. 2023. "Stock Price Prediction Using Machine Learning." In *2023 International Conference on Computational Intelligence and Sustainable Engineering Solutions (CISES)*, pp. 926–932. IEEE.

Ma, Shaojun; Li, Lei, Ke, Huimin, and Yilin, Zheng. 2022. "Environmental Protection, Industrial Structure and Urbanization: Spatiotemporal Evidence from Beijing–Tianjin–Hebei." *China Sustainability*14(2). p. 795. https://doi.org/10.3390/su14020795

Mitchell, A.S. 2019. A Smarter Route to Smart Cities. KPMG [Online]. https://home.kpmg/xx/en/home/insights/2018/04/a-smarter-route-to-smart-cities.html

National Urban Policy Framework. 2018. Ministry of Housing & Urban Affairs [Online]. https://smartnet.niua.org/sites/default/files/resources/nupf_final.pdf

O'Brien, K. 2018. From Smart City to Future City: The 21st-Century Urbanization Challenge [Online]. https://www.digitalistmag.com/improving-lives/2018/01/26/from-smart-city-to-future-city-21st-century-urbanization-challenge-05809136

Parmar, M. (Ed.). 2017. *Reflections*. Mumbai. KRVIA Publications Cell, Kamla Raheja Vidyanidhi Institute for Architecture and Environmental Studies.

Prashar, N., Hooda, S., and Kumar, R. 2023. Cybersecurity and Evolutionary Data Engineering: Current Status of Challenges in Data Security: A Review. pp (3–13). https://doi.org/10.1007/978-981-99-5080-5_1

Saunders, Tom and Baeck, Peter. 2015. Rethinking Cities from the Ground Up. https://ofti.org/wpcontent/uploads/2015/06/rethinking_smart_cities_from_the_ground_up_2015.pdf

Schuurman, D., Marez, L., and Ballon, P. 2016. "The Impact of Living Lab Methodology on Open Innovation Contributions and Outcomes." *Technology Innovation Management Review*, 6(1), 7–16. https://doi.org/10.22215/timreview/956

Seunghwan, Myeong, Jaehyun, Park, and Lee, Minhyung. 2022. "Research Models and Methodologies on the Smart City: A Systematic Literature Review." *Sustainability*. https://doi.org/10.3390/su14031687

Singh, B. and Parmar, M. 2019. *Smart City in India: Urban Laboratory, Paradigm or Trajectory?*. India. Routledge.

Smart Sustainable Cities. 2015. UNECE and ITU[Online]. https://www.itu.int/en/ITU-T/ssc/united/Pages/default.aspx

Sung-Han, Sim, and Lee, Jong-Jae. 2021. "Special Issue on Smart City and Smart Infrastructure." *Sensors*. https://doi.org/10.3390/S21217064

Telecommunication Engineering Center. 2019. "Design and Planning Smart Cities with IoT/ICTTEC Report." Ministry of Communications, Government of India. http://tec.gov.in/pdf/M2M/Design%20Planning%20Smart%20Cities%20with%20IoT%20ICT.pdf

Zhao, Z. and Zhang, Y. 2019. Impact of Smart City Planning and Construction on Economic and Social Benefits Based on Big Data Analysis. *Complexity*, 2020(1), 8879132. https://doi.org/10.1155/2020/8879132

4 Different Methodologies for Smart City and Urbanization System

Kamal Kumar, Poonam, Rupinder Kaur and Varsha Rani

Ch. Devi Lal State Institute of Engineering & Technology, Sirsa, India

4.1 INTRODUCTION

Growing interest in "smart cities," which use connectivity, data, and technology to boost productivity, sustainability, and quality of life, is a result of the rapid urbanization trend. Smart cities are defined differently, but they generally combine big data analytics, Internet of Things (IOT) sensors, and other advances with information and communication technology (ICT) in many urban areas (Kang, Lei, and Siyou, Xia, 2023). This enables intelligent management of assets and provision of services. Areas impacted include energy, water, waste, mobility, healthcare, public safety and more.

Smart city development aims to address pressing urban challenges like climate change, resource constraints, congestion, pollution, poverty and aging infrastructure. Montreal Smart and Digital City Methodology has been shown in Figure 4.1. The integration of digital technologies provides the visibility and control needed to optimize systems (This enhances economic prosperity and sustainability). Citizens also benefit through improved access to services (Ma, Shaojun, Li, Lei, Ke, Huimin, and Yilin, Zheng, 2022), (Sim and Lee, 2021) *United 4 Smart Sustainable Cities, UNECE and ITU* [2025].

However, smart city implementations face complex technical, governance, social and adoption barriers (Sim and Lee, 2021) (Hooda et al., 2021, 2023). Substantial investments are required for sensors, broadband, cloud platforms, analytics, network security and skills development (Anthopoulos, L.G., Janssen, M. and Weerakkody, 2016). Engaging diverse stakeholders across public, private and community sectors is difficult (Bakıcı, T., Almirall, E. and Wareham, 2013). Many pilots struggle to scale or generate returns (Batty, M., Axhausen, K. W., Giannotti, F., Pozdnoukhov, A., Bazzani, A., Wachowicz, M., and Portugali, Y., 2012). Still, the smart city market continues to grow as urbanization accelerates globally.

This article provides a comprehensive overview of smart city research and practice as shown in Figure 4.2. Section II surveys definitions and frameworks. Section III analyses key application domains. Section IV examines emerging technologies and roles. Section V presents real-world case studies. Section VI discusses challenges. Section VII provides conclusions and future outlook.

DOI: 10.1201/9781003467892-4

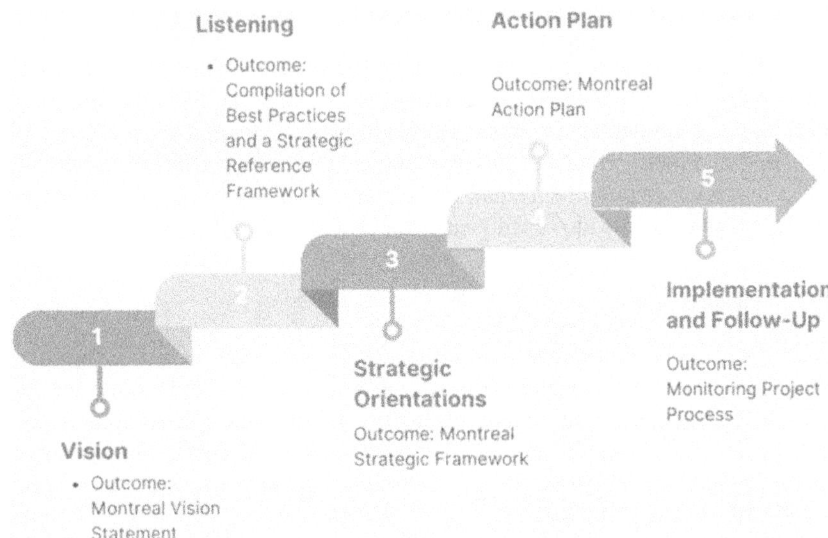

FIGURE 4.1 Montreal Smart and Digital City methodology.

(Source: Montreal Smart and Digital City).

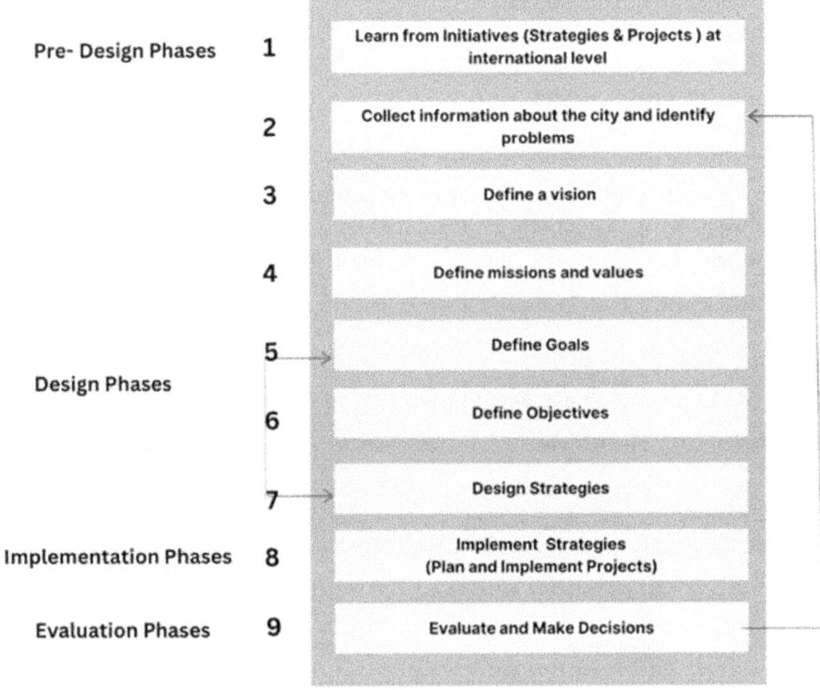

FIGURE 4.2 Strategy development process.

(Source: mdpi).

4.2   SMART CITY DEFINITIONS AND FRAMEWORKS

There is no universally accepted definition of a smart city (Batty, M., Axhausen, K. W., Giannotti, F., Pozdnoukhov, A., Bazzani, A., Wachowicz, M., and Portugali, Y., 2012). Conceptualizations emphasize different aspects, though most highlight the use of ICT and data to enhance city operations and services (Sim and Lee, 2021). This section reviews prominent smart city definitions and syntheses key dimensions. Leading frameworks are also examined.

4.2.1   DEFINITIONS

In Table 4.1, Initial definitions focused on technology, such as using smart computing to improve infrastructure and services (Sim and Lee, 2021). Later meanings highlighted intelligence, sustainability, quality of life and empowering residents (Anthopoulos, L.G., Janssen, M. and Weerakkody, 2016). Recent definitions increasingly recognize the importance of collaborative governance and human-centric design (Sim and Lee, 2021).

Analysis of numerous smart city meanings reveals several core aspects (Saunders and Baeck, 2015):

- Use of ICT, IoT, big data and analytics.
- Networked infrastructure and integration.
- Sustainability and resilience.
- Innovation and competitiveness.
- Citizen engagement and inclusion.
- Enhanced quality of life.

While specific technocentric, ecocentric or anthropocentric orientations persist, contemporary thinking embraces smart cities as complex socio-technical systems (Batty, M., Axhausen, K. W., Giannotti, F., Pozdnoukhov, A., Bazzani, A., Wachowicz, M., and Portugali, Y., 2012). This balances technology with governance, people and

---

TABLE 4.1

Project Management Phase Breakdown (Based on Saunders and Baeck, 2015)

Phase Number	Phase Name	Step Number	Step Name
1	Context Analysis	1	Benchmarking Best Practices
		2	Current State Analysis
		3	Smart City Stakeholders
		4	Institutional Capacity
2	Strategy	5	Smart City Vision
		6	Objectives and Governance Model
3	Action Plan	7	Align Initiatives with Smart City Domains and SDGs
		8	Piloting Projects and Programmes
4	Evaluation	9	Project Monitoring
		10	Impact Reporting

policy (Saunders and Baeck, 2015). Still, ambiguity around the smart city concept remains (Sim and Lee, 2021). Further synthesis is needed to reconcile perspectives (Saunders and Baeck, 2015).

4.2.2 FRAMEWORKS

Many reference frameworks describe smart city dimensions and components as shown in Figure 4.3. While varying in structure, emphases and details, common high-level elements emerge. Six foundational smart city pillars are summarized below:

1. *Technology and Innovation*: Deployment of ICT infrastructure (broadband, cloud, sensors, networks), Internet of Things, big data platforms, and emerging technologies to enable real-time monitoring, optimization and service improvements (Batty, M., Axhausen, K. W., Giannotti, F., Pozdnoukhov, A., Bazzani, A., Wachowicz, M., and Portugali, Y., 2012).
2. *Sustainability and Resilience*: Holistic planning and operations to reduce environmental impact, resource use, pollution and emissions while adapting to climate change.
3. *Governance and Policy*: New models of government leadership, multi-stakeholder collaboration, digital administration, innovation promotion and regulatory support for smart cities (Batty, M., Axhausen, K. W., Giannotti, F., Pozdnoukhov, A., Bazzani, A., Wachowicz, M., and Portugali, Y., 2012).
4. *Economy and Infrastructure*: Strategic investments in physical infrastructure (energy, water, waste, transport), digital connectivity, and economic growth and job creation (Bakıcı, T., Almirall, E. and Wareham, 2013).

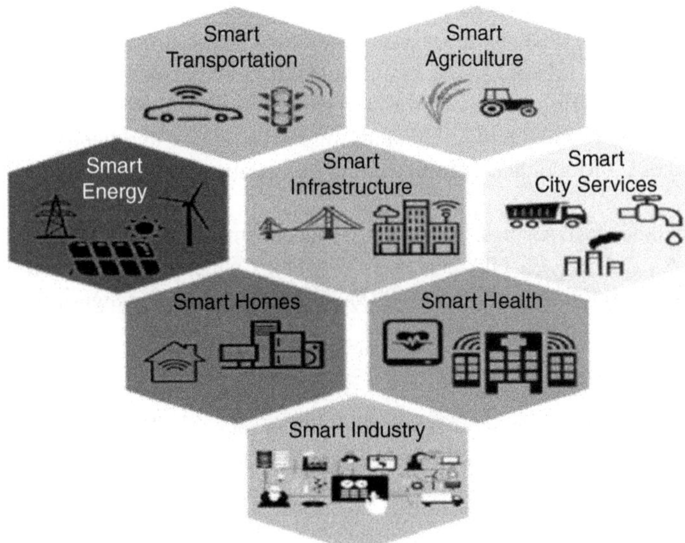

FIGURE 4.3 Conceptualizations of smart city governance.

(Source: FG-SCC Technical report on Shareholders for Smart Sustainable Cities).

5. *People and Communities*: Citizen participation, digital empowerment and inclusion, personalized services, education, social innovation and improved quality of life (Anthopoulos, L.G., Janssen, M. and Weerakkody, 2016).
6. *Data and Analytics*: Management of vast integrated data resources to generate real-time operational insights and guide evidence-based planning and decision-making (Batty, M., Axhausen, K. W., Giannotti, F., Pozdnoukhov, A., Bazzani, A., Wachowicz, M., and Portugali, Y., 2012).
7. These pillars are interdependent and mutually reinforcing. Successful smart city development requires coordinated progress across dimensions supported by enabling policy, finance and partnerships. This drives an integrated vision.

4.3   KEY APPLICATION DOMAINS

Smart city solutions span diverse urban sectors (Batty, M., Axhausen, K. W., Giannotti, F., Pozdnoukhov, A., Bazzani, A., Wachowicz, M., and Portugali, Y., 2012). This section analyses high-priority domains including energy, mobility, government, infrastructure, healthcare and public safety.

4.3.1   SMART ENERGY

Energy systems are a major focus of smart city innovation (Caragliu, A., Del Bo, C., Nijkamp and P., 2011). Intelligent grids, meters, management and efficiency technologies are deployed. Key applications include:

- Smart metering – Collects granular consumption data, enables monitoring and control.
- Smart grids – Integrate IT to balance supply and demand, reduce outages.
- Microgrids – Localized grids with distributed generation and storage.
- Renewables – Expand solar, wind, biomass; integrate with grid.
- Efficiency – Minimize usage through smart building controls.
- Electric vehicles – Deploy EV charging infrastructure and fleets.
- Analytics – Model and optimize complex energy ecosystem.

Benefits include cost savings, reliability, reduced emissions and climate resilience (Caragliu, A., Del Bo, C., Nijkamp and P., 2011). Challenges include cybersecurity, interoperability, and adoption (Gruber, 1993). Still, smart energy is a high-impact domain.

4.3.2   SMART MOBILITY

Transportation systems are another priority (Chourabi et al., 2012). Solutions aim to reduce congestion, improve traffic flow, enhance road safety and provide intelligent mobility services. Smart Intersection availability map has been shown in Figure 4.4. Examples include:

- Intelligent Transport Systems (ITS) – Apply communication, control, and sensing.
- Traffic monitoring – Use cameras, sensors, video analytics to analyse flow.

FIGURE 4.4   Smart intersection availability map.

(Source: Urfalı and Eymen, 2020).

- Traffic optimization – Adapt signals based on real-time conditions.
- Autonomous vehicles – Develop self-driving technology and infrastructure.
- Multimodal integration – Unify trip planning and payments across transport.
- Mobility services – Offer shared mobility, ride-hailing, microtransit options.
- Analytics – Model travel patterns and mobility dynamics.

Benefits include accessibility, sustainability, economic productivity and urban liveability (Hollands, R. G., 2008). Barriers include cost, safety, jobs and adoption (Gruber, 1993). Smart mobility is a complex challenge.

4.3.3 SMART GOVERNANCE

Governments play central roles in smart city leadership, planning and coordination (Borst, 1997). Figure 4.5 shows government initiatives to improve public administration and services using:

- Digital workflows – Automate and optimize back-end processes.
- Integrated data – Consolidate information across departments.
- Analytics – Extract insights to guide decisions and policies.
- Cloud services – Virtualize infrastructure to enable flexibility.
- Mobile access – Provide real-time self-service via apps and portals.
- Smart sensors – Monitor urban environment and public assets.
- Open data – Promote innovation and transparency with public data.

This enhances efficiency, transparency, collaboration and policy intelligence (Hollands and R. G., 2008). Barriers include cybersecurity, legacy systems and adoption. Smart governance establishes a digital foundation.

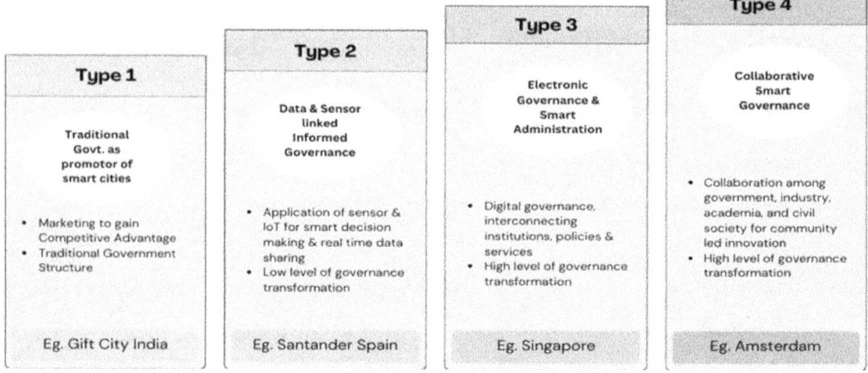

FIGURE 4.5   The Role of eGovernment in Smart City Construction.

(Source: Praharaj, Han and Hawken, 2018).

4.3.4 Smart Infrastructure

Physical infrastructure must be made intelligent to support smart cities (Komninos, N., Pallot, M., Schaffers and H, 2013). This involves:

- Smart buildings – Incorporate automation, control and optimization.
- Water management – Use smart meters, leak detection, quality monitoring.
- Waste management – Implement sensorized bins, route optimization, analytics.
- Environmental monitoring – Deploy dense real-time air/water quality sensing.
- Public lighting – Use intelligent, energy-efficient lighting systems.
- Digital twins – Create virtual models of infrastructure for management.

Benefits include efficiency, sustainability and improved services. Cost and integration challenges remain. Modernizing infrastructure is imperative.

4.3.5 Smart Healthcare

Healthcare is both a major social need and cost factor in cities that can be improved through smart approaches (Nam and Pardo, 2011a and 2011b):

- Telemedicine – Provide remote clinical services and virtual visits.
- mHealth – Enable patient monitoring, appointments and records via mobile.
- Health sensing – Collect real-time population health data with wearables.
- Digital health records – Integrate electronic records across providers.
- Spatial analytics – Map population health patterns and clinical needs.
- AI diagnosis – Apply predictive algorithms to support clinicians.

This expands access, improves outcomes and reduces expenses. Privacy and integration are ongoing concerns. Better health and wellness are essential.

4.3.6　PUBLIC SAFETY AND EMERGENCY RESPONSE

Smart technologies can also enhance urban safety and emergency preparedness (Neirotti, P., De Marco, A., Cagliano, A. C., Mangano, G., and Scorrano, F., 2014):

- Crime mapping and prediction – Understand incidents and deploy resources efficiently.
- Crowd monitoring – Detect congestion and disasters with video analytics.
- Multimodal emergency communication – Leverage various channels to reach citizens.
- Intelligent disaster response – Integrate real-time information for coordination.
- Next-generation 911 – Incorporate multimedia and data into emergency systems.

Benefits include rapid response, improved services and resilience. Adoption, accuracy and budgets can constrain impact (Neirotti, P., De Marco, A., Cagliano, A. C., Mangano, G., and Scorrano, F., 2014). Public safety is a priority application.

4.4　EMERGING TECHNOLOGIES AND ROLES

Smart cities depend on innovation across the technology stack, from devices to networks to analytics (Washburn, D., Sindhu, U., Balaouras, S., Dines, R. A., Hayes, N. M., and Nelson, L. E. 2010). This section examines key emerging technologies and their smart city roles as shown in Figure 4.6.

4.4.1　5G AND CONNECTIVITY

High-bandwidth low-latency 5G networks will enable next-generation smart city solutions (Washburn, D., Sindhu, U., Balaouras, S., Dines, R. A., Hayes, N. M. and Nelson, L. E., 2010). Enhanced capacity, speed, reliability and connection density empower applications like:

- Real-time video security and analytics.
- Autonomous industrial and vehicular control.
- Augmented and virtual reality experiences.
- Connected health and telemedicine.
- Multimedia smart services.
- Massive IoT device management.

5G unlocks new realms of live data-driven urban management. Rollouts are accelerating globally (Studer, Benjamins, andFensel,1998). Networks are a smart city foundation.

4.4.2　ARTIFICIAL INTELLIGENCE

AI will enhance automated decision-making across smart city domains.

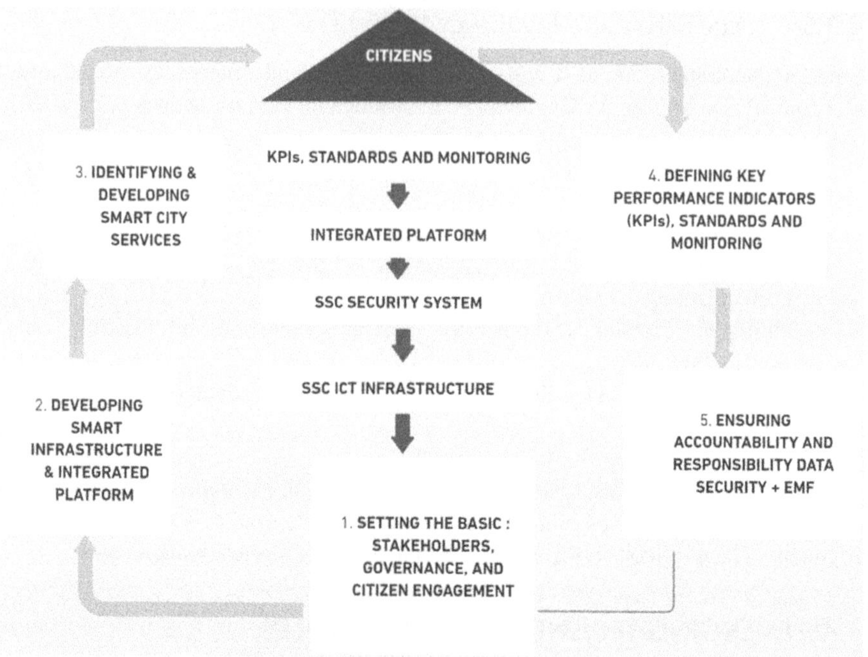

FIGURE 4.6   Emerging technologies and their smart city Roles.

(Source: Based on Taamallah, Khemaja and Faiz, 2019).

Key techniques include:

- Machine learning – Systems self-improve via data patterns.
- Deep learning – Advanced neural networks uncover complex insights.
- Natural language processing – Let systems parse human speech and text.
- Computer vision – Algorithms analyze images and video.
- Predictive analytics – Forecast outcomes and recommend actions.
- Robotics – Enable intelligent physical assistance and automation.

AI can make operations more efficient, personalized and far-reaching (Studer, Benjamins, andFensel,1998). Careful design is needed to ensure proper governance (Gruber, 1993). AI-enabled intelligence is vital.

4.4.3   INTERNET OF THINGS

Networked physical devices and sensors will expand smart city data collection (Gruber, 1993). Major applications:

- Infrastructure monitoring – Track performance metrics of roads, bridges, rail, water systems, etc.
- Environmental sensing – Distributed measurement of air/water quality, pollution, noise.

- Resource management – Meter energy, water consumption; manage waste.
- Condition monitoring – Assess attributes like crowd levels, parking availability.
- Logistics monitoring – Track vehicle locations and shipments.
- Health monitoring – Wearables to collect patient vital signs in realtime.

Advanced IoT systems will provide comprehensive real-time city visibility. Interoperability and scale are ongoing obstacles (Borst, 1997). IoT is integral to smart city development.

4.4.4 Big Data and Analytics

Vast data from sensors, meters, cameras, vehicles, social media and institutional records requires big data infrastructure to store, process and analyze (Borst, 1997). Sophisticated analytics can uncover:

- Operational insights – Optimize performance and resource allocation.
- Infrastructure patterns – Predict maintenance needs and failures.
- Mobility models – Manage congestion and transit capacity.
- Disease outbreaks – Enable public health response through symptom monitoring.
- Citizen perceptions – Gauge public opinion and engagement.
- Environmental trends – Track pollution and climate impacts.
- Socioeconomic indicators – Inform policy for employment, equity, development, etc.

Advanced simulation, machine learning and AI techniques enable sophisticated modelling Private data protection is imperative (Mario Piattini, Coral Calero, and Francisco Ruiz, 2006). Big data analytics unlock smart city intelligence.

4.4.5 Blockchain

Shared distributed ledger technology can enhance smart city operations and services (Mario Piattini, Coral Calero, and Francisco Ruiz, 2006):

- Smart contracts – Execute actions based on trusted data triggers.
- Digital identity – Securely manage credentials for authentication.
- Energy trading – Record and validate peer-to-peer energy transactions.
- Supply chain – Track integrity of goods and provenance records.
- Voting – Implement transparent and tamper-resistant voting.
- Data marketplace – Enable trusted data monetization and sharing.

Blockchain builds smart city trust and resilience (Pisanelli, Gangemi, and Steve, 2002). Scalability limits remain a constraint. The technology holds long-term potential.

4.4.6 Cloud Computing

Cloud platforms provide storage, computing and analytics resources to power smart city IT (Rana, N.P., Luthra, S., Mangla, S.K., Islam, R., Roderick, S., and Dwivedi Y.K., 2019). Benefits:

- On-demand capacity – Scale based on dynamic needs.
- Speed of deployment – Provision resources quickly.
- Pay-as-you-go – Consume and pay only for what is used.
- Analytics access – Leverage high-power cloud-based data services.
- Resilience – Leverage distributed infrastructure with failover.
- Accessibility – Provide universal access to applications and data.

Cloud delivers agile, scalable and flexible smart city IT (Becerril-Elías, J.C., Merritt, H. and Alianzas., 2021). Security and availability risks persist (X. Wang, C. Chan, and H. Hamilton, 2002). Cloud is a vital smart city enabler.

4.5 SMART CITY CASE STUDIES

This section analyses real-world smart city implementations in different regions. Diverse examples provide practical insights into emerging models and leading practices.

4.5.1 PlanIT Valley, Portugal

PlanIT Valley is a new community built from the ground up as an experimental smart city testbed (Korczak, J., Kijewska and K. 2019). Located in Porto, Portugal, the 125-hectare site applies a top-down smart city approach led by private real estate and technology partners, including Living PlanIT. Energy, water, transport and other urban systems are digitally instrumented, monitored and optimized. A cloud platform integrates sensor data to enable intelligent management. The greenfield deployment provides a controlled environment to pilot new concepts before adapting solutions to existing cities. PlanIT Valley demonstrates comprehensive technocentric smart city development.

4.5.2 Songdo, South Korea

Songdo, near Seoul, exemplifies new purpose-built smart cities in Asia (Purvis, B., Mao, Y., Robinson and D., 2019). The $35 billion project focuses on advanced networking, data analytics and sustainable design. Songdo applies a systems integration methodology with centralized management of municipal operations including energy, water, waste and transportation. A ubiquitous connectivity blanket provides the foundation for smart services delivered via smartphones. The city is built on a sustainable footprint, including 40% green space. Songdo has emerged as a benchmark for planned Asian ecocities.

4.5.3 Santander, Spain

Santander, Spain pioneered city-scale IoT deployments under the European Union SmartSantander initiative OECD [2017]. The project has embedded over 12,000 sensors throughout the city to monitor parking spaces, traffic, street lighting, waste management and environmental conditions. Sensor platforms were tested by both the municipality and private companies to support applications from traffic pattern analysis to irrigation optimization. Santander provided an important real-world validation of massive urban sensor network potential.

4.5.4 Barcelona, Spain

Barcelona is a leader in technology-enabled citizen engagement and social innovation (Mensah and J., 2019). Urban interaction design concepts are showcased at the Institute for Advanced Architecture of Catalonia. City governance promotes grassroots digital fabrication labs (FabLabs), smart citizen participation platforms, and community knowledge exchange. Barcelona applies a unique human-centric smart city paradigm focused on the cultivation of inclusive digital culture. This flips the top-down technocratic model to empower citizens.

4.5.5 Medellin, Colombia

Medellin provides lessons in smart city social inclusion (Öberg, C., Graham, G., Hennelly and P, 2017). Technology investments specifically target the poorest marginalized districts to provide libraries, schools, transit and community facilities. This helps bridge the divide, building trust and expanding opportunity. Medellin also promotes environmental sustainability through extensive green corridors and public spaces. The city's metropolitan area office brings a unique regional dimension as well. Medellin demonstrates smart cities placing priority on people.

4.5.6 Shanghai, China

Shanghai aims to be a global exemplar of New Smart Cities blending intelligence, sustainability and quality of life (Caragliu, A., del Bo, C.,Nijkamp, and P, 2011). The Shanghai Tower incorporates extensive automation and green design. Investments in wireless broadband, IoT and cloud computing enable big data-driven operations optimization. The Municipal Government leverages technology to coordinate agencies, support startups and improve services. Major smart transportation pilots are underway as well. Public–private partnerships help drive rapid progress. Shanghai shows comprehensive top-down smart city development.

4.5.7 Mumbai, India

Mumbai applies smart city innovations to address rapid growth and strained public services (Antrobus and D. 2011). Plans include intelligent traffic management, water distribution, waste collection and disaster response systems. Digital empowerment initiatives bridge the digital divide through free WiFi, mobile apps for civic services and open data portals. Despite major infrastructure and economic challenges, Mumbai makes steady progress in implementing targeted smart city solutions with social impact.

4.5.8 Singapore

Singapore's 'smart nation' initiative provides integrated governance, business and technology leadership (Patton and M.Q, 2005), (Izquierdo and G.M. 2015, Martínez, I.P., Alvareza and R.M., 2019). The government aggressively invests in digital infrastructure like broadband and wireless networks. Data exchange platforms and analytics centres enable collaborative applications spanning health, education, transport and finance. Regulatory policies adaptively support innovation. Led by the Prime Minister's

> Copenhagen, Denmark – Leader in smart city services and citizen engagement through platforms like City Data Exchange and City SDK. Strong focus on sustainability and quality of life.
> London, UK – Applies extensive sensor networks, data analytics, and operating centres to manage transport, emergency response, utilities and governance. Open data initiatives spur innovation.
> Boston, USA – Leverages MIT expertise to pilot smart city projects including intelligent parking meters, autonomous vehicles, and city data analytics. Public–private partnerships drive progress.

4.5.8.1   Smart Intersections

Smart intersections optimize traffic flow, reduce waiting times, provide environmental benefits, support the economy, and enhance safety. Specifically, they improve traffic movement through intersections by adapting signals based on real-time conditions, preventing unnecessary vehicle waits and lost time. This accelerates travel, lessens fuel waste and emissions, boosts productivity from less congestion, and decreases accidents via cautious signalling.

The technology works by using advanced systems to control lights, counting vehicles with cameras and sensors to gather data, and automatically adjusting green light duration based on approaching vehicle density. By determining which intersection branches have more cars, priority and flow is directed effectively. The integrated data and responsive signals create maximized travel efficiency.

4.5.8.2   Smart Intersections: Behind the Scenes

Like an orchestra conductor, smart intersections use advanced systems, sensors, and data to synchronize traffic flow. The integrated technology works in harmony to create transportation efficiency through responsive signals and optimized routes.

The foundation relies on intelligent signalling controllers running advanced software at the intersection. These adaptive systems ingest real-time data to direct traffic like a rail yard dispatcher – continuously adjusting signals to keep vehicles rolling based on changing conditions.

Feeding the data streams are two key instrument sections – cameras and sensors. Vehicle counting cameras utilize visual processing algorithms to monitor approaching traffic density on different arms. Buried road sensors also track vehicle flows and relay metrics to the conductor system.

The conductor software harmonizes the camera and sensor inputs to optimize light patterns in an instant, like improvising a new passage. If more cars queue on the northbound lane, green lights adapt to give them priority, preventing pile-ups. The system fluidly shifts priorities as vehicles ebb and flow.

With technology now allowing orchestrated signals to match real-time conditions, smart intersections form a well-tuned ensemble keeping the urban commute flowing smoothly. The result is less congestion, reduced wasted time, and smarter mobility.

Overall, smart intersections use modern data collection and adaptive signalling to ease congestion and facilitate smarter, safer transportation for both drivers and pedestrians. The optimization cuts down on travel delays, pollution, and crashes while improving the economy.

In conclusion, smart intersections are a promising solution to address the challenges posed by increasing urbanization and the growth of vehicles on the road. By optimizing traffic signal timings and utilizing real-time data, smart intersections contribute to improved traffic flow, reduced environmental impact, enhanced safety, and economic benefits for the country (Figure 4.7).

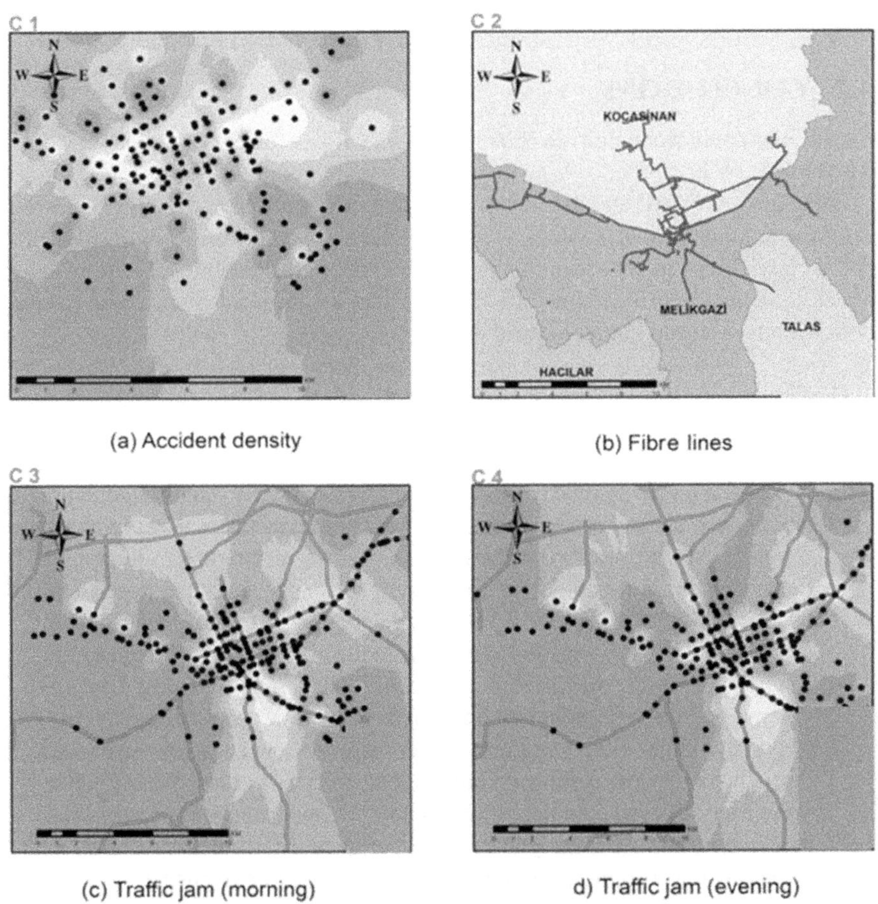

FIGURE 4.7 Reclassification maps used for four sub-criteria**. (a) Number of accidents at Intersections (b) Distance of fibre lines (c) Number of vehicles in intersections (morning) (d) Number of vehicles in intersections (evening).

(Source: Urfalı and Eymen, 2020).

4.6 CHALLENGES

- Costs – Smart city technology requires major investments in sensors, broadband, platforms, network security and skills.
- Technology integration – Diverse systems must be unified. Interoperability, standards, architecture design are key issues.
- Data management – Enormous data volumes require policies, infrastructure, analytics skills to provide value.
- Cyber security – Critical infrastructure and data systems must be secured against growing threats.
- Privacy – Personal data collection and use raise privacy concerns requiring safeguards.
- Governance – Leadership vision, cross-sector collaboration, and policy innovation are essential.

4.7 CONCLUSIONS

Smart cities apply technology to enhance sustainability, efficiency, services and quality of life.

Integrated ICT, IoT, and data drive intelligent management and decision-making. Progress requires holistic strategies spanning technology, infrastructure, governance and society. Pillars include technology and innovation, sustainability, economy, governance, society and data analytics. Success factors are leadership, collaboration, policy innovation and human-centric design.

1. Technology and Innovation: Smart cities harness cutting-edge technology such as 5G networks, artificial intelligence (AI), and blockchain to optimize resource utilization.

   Innovation centres and technology hubs foster creativity and collaboration among startups and established tech companies. Investment in research and development encourages the continuous evolution of smart city solutions.

2. Sustainability: Renewable energy sources like solar and wind power contribute to a greener and more sustainable energy grid. Waste management systems employ IoT sensors to optimize collection routes and reduce environmental impact. Sustainable architecture and urban planning prioritize green spaces, eco-friendly buildings, and reduced carbon emissions.

3. Economy: Smart cities stimulate economic growth by attracting businesses through tax incentives and streamlined regulatory processes. The digitalization of services and commerce creates new job opportunities in sectors such as e-commerce, data analytics, and cyber security. Entrepreneurship and small business support programs foster local economic resilience.

4. Governance: Transparent and responsive governance is vital in smart cities, with open data initiatives and public participation platforms. City administrations use data-driven decision-making to allocate resources efficiently and address citizen needs. Collaboration between government agencies, private sector partners, and citizens ensures a well-coordinated approach to city management.

5. Society: Inclusive policies and programs prioritize social equity, ensuring that all citizens benefit from smart city initiatives. Smart education and healthcare systems enhance access to quality services, regardless of socio-economic background. Community engagement initiatives empower citizens to actively participate in shaping their city's future.

6. Data Analytics: Advanced data analytics tools process vast amounts of data generated by IoT sensors, enabling predictive maintenance, traffic optimization and more.

   Data privacy and security measures are in place to protect citizens' information and ensure ethical data usage.

   Real-time data analytics provide insights that help city leaders make informed decisions.

7. Success Factors:

   Leadership: Strong visionary leadership is essential to drive smart city initiatives, set clear goals and inspire collaboration.

   Collaboration: Partnerships between public and private sectors, academia and civil society facilitate knowledge sharing and innovation.

   Policy Innovation: Agile and forward-thinking policies adapt to the rapidly changing technology landscape and regulatory challenges.

   Human-Centric Design: Smart city solutions prioritize the well-being and needs of residents, ensuring that technology serves people, not the other way around.

BIBLIOGRAPHY

Albino, V., Berardi, U., Dangelico, R.M. 2015. Smart Cities: Definitions, Dimensions, Performance, and Initiatives. *Journal of Urban Technology*, Vol. 22, 3–21.

Alderete, M.V. 2019. QuéFactoresInfluyen En La Construcción de CiudadesInteligentes? Un ModeloMultinivel Con Datos a Nivel Ciudades y Países. *Revista Iberoamericana de Ciencia, Tecnología y Sociedad-CTS*, Vol. 14, 71–89.

Anthopoulos, L., Fitsilis, P. 2010. From Digital to Ubiquitous Cities: Defining a Common Architecture for Urban Development. *Proceedings of the 2010 6th International Conference on Intelligent Environments (IE 2010)*. Kuala Lumpur, Malaysia. pp. 301–306.

Anthopoulos, L.G., Janssen, M., Weerakkody, V. 2016. Smart Service Portfolios: Do the Cities Follow Standards? *Government Information Quarterly*, Vol. 33(4), 663–672.

Antrobus, D. 2011. Smart Green Cities: From Modernization to Resilience? *Urban Research & Practice*, Vol. 4, 207–214.

Bai, X., Nath, I., Capon, A., Hasan, N., Jaron, D. 2012. Health and Wellbeing in the Changing Urban Environment: Complex Challenges, Scientific Responses, and the Way Forward. *Current Opinion in Environmental Sustainability*, Vol. 4, 465–472.

Bakıcı, T., Almirall, E., Wareham, J. 2013. A Smart City Initiative: The Case of Barcelona. *Journal of the Knowledge Economy*, Vol. 4(2), 135–148.

Batty, M., Axhausen, K. W., Giannotti, F., Pozdnoukhov, A., Bazzani, A., Wachowicz, M., Portugali, Y. 2012. Smart Cities of the Future. *The European Physical Journal Special Topics*, Vol. 214(1), 481–518.

Becerril-Elías, J.C., Merritt, H. A. 2021. para la innovaciónenorganizacionesintensivasenconocimiento: El caso de México. *Revista CEA*, Vol. 7, e1780.

Belissent, J., Giron, F. 2013. *Service Providers Accelerate Smart City Projects: Forrester Research Report*. London, UK. Forrester Publication United Kingdom. https://silo.tips/download/service-providers-accelerate-smart-city-projects

Bonham-Carter, C., Clark, W.W. 2009. *Sustainable Communities*. 1st ed. New York, USA, Springer.

Borst, W.N. 1997. Ph.D. Thesis. Construction of Engineering Ontologies.

Bourkela, M., Casseb, M., Bassi, S., De Lucca, C., Facchina, M. (Eds.). 2016. *The Road Toward Smart Cities*. Inter—American Development Bank (IDB). https://publications. iadb.org/publications/english/document/The-Road-toward-Smart-Cities-Migrating-from-Traditional-City-Management-to-the-Smart-City.pdf

Caragliu, A., del Bo, C. and Nijkamp. 2011. Smart Cities in Europe. *Journal of Urban Technology*, Vol. 18, 65–82.

CASBEE. 2014. *CASBEE for Urban Development: Comprehensive Assessment System for Building Environmental Efficiency*. 2014 ed. Technical Manual. https://www.ibec.or.jp/ CASBEE/english/

City of Montréal. 2014. *Smart and digital city - 2014-2017 Montréal strategy* (pp. 15–40). Bureau de la ville intelligente et numérique de Montréal.

Chourabi, H., Nam, T., Walker, S., Gil-Garcia, J.R., Mellouli, S., Nahon, K., Pardo, T.A. and Scholl, H.J. 2012. Understanding Smart Cities: An Integrative Framework. *Proceedings of the Annual Hawaii International Conference on System Sciences IEEE Computer Society*. Maui, HI, USA. 4–7; pp. 2289–2297.

De Guimarães, J.C.F., Severo, E.A., Felix Júnior, L.A., Da Costa, W.P.L.B. and Salmoria, F.T. 2020. Governance and Quality of Life in Smart Cities: Towards Sustainable Development Goals. *Journal of Cleaner Production*, Vol. 253, 119926.

Feil, A.A. and Schreiber, D. 2017. Sustentabilidade e DesenvolvimentoSustentável: Desvendando as Sobreposições e Alcances de Seus Significados. *Cadernos Ebape. BR*, Vol. 15, 667–681.

Fischer, L.-B. and Newig, J. 2016. Importance of Actors and Agency in Sustainability Transitions: A Systematic Exploration of the *Literature. Sustainability*, Vol. 8, 476.

Gandrita, D. M. 2023. Improving Strategic Planning: The Crucial Role of Enhancing Relationships Between Management Levels. *Administrative Sciences*, 13(10), 211. https://doi.org/10.3390/admsci13100211

Gruber, T.R. 1993. A Translation Approach to Portable Ontology Specifications. *Knowledge Acquisition*, Vol. 5(2), 199–221.

Harrison, C., Eckman, B., Hamilton, R., Hartswick, P., Kalagnanam, J., Paraszczak, J. and Williams, P. 2010. Foundations for Smarter Cities. *IBM Journal of Research and Development*, Vol. 54, 1–16.

Heymans, S., et al. 2008. *Ontology Reasoning with Large Data Repositories*. Vol. 7. Ontology Management- Computing for Human Experience.

Hollands, R. G. 2008. Will the Real Smart City Please Stand Up? Intelligent, Progressive or Entrepreneurial? *City*, Vol. 12(3), 303–320.

Hooda, Susheela. 2021. A New Approach for Power-Aware Routing for Mobile Adhoc Networks Using Cluster Head With Gateway Table. *International Journal of Web-Based Learning and Teaching Technologies (IJWLTT)*, Vol. 16, 47–59. https://doi.org/10.4018/ IJWLTT.20210701.oa4

Hooda, Susheela, Lamba, Vikas, and Kaur, Amandeep. 2021. AI and Soft Computing Techniques for Securing Cloud and Edge Computing: A Systematic Review, Paper presented at *5th International Conference on Information Systems and Computer Networks (ISCON)*, Mathura, India, pp. 1–5. https://doi.org/10.1109/ISCON52037.2021.9702422

Horrocks, Ian, and Sattler, Uli. 2001. Ontology Reasoning in the SHOQ(D) Description Logic. *International Joint Conference on Artificial Intelligence*. Vol. 1.

Indian Green Building Council. 2010. *IGBC Green Townships Rating System*. Abridged Reference Guide Pilot Version Ed. Telangana. India Indian Green Building Council.

Ismagilova, E., Hughes, L., Dwivedi, Y.K. and Raman, K.R. 2019. Smart Cities: Advances in Research—An Information Systems Perspective. *International Journal of Information Management*, Vol. 47, 88–100.

Izquierdo, G.M. 2015. Informantes y Muestreo En Investigación Cualitativa. *Investigaciones andina*, Vol. 17, 1148–1150.

Kang, Lei, and Siyou, Xia. 2023. Study on Urbanization Sustainability of Xinjiang in China: Connotation, Indicators and Measurement. *International Journal of Environmental Research and Public Health*, Vol. 20(3), 2535. https://doi.org/10.3390/ijerph20032535

Kang, Yong Bin, Shonali Krishnaswamy, Wudhichart Sawangphol, Lianli Gao, and Yuan-Fang Li. 2019. Understanding and Improving Ontology Reasoning Efficiency Through Learning and Ranking. *Information Systems*, Vol. 87.

Kaur, H., and Garg, P. 2019. Urban Sustainability Assessment Tools: A Review. *Journal of Cleaner Production*, Vol. 210, 146–158.

Klarin, T. 2018. The Concept of Sustainable Development: From its Beginning to the Contemporary Issues. *ZIREB Zagreb International Review of Economics and Business*, Vol. 21, 67–94.

Komninos, N., Pallot, M., and Schaffers, H. 2013. Special Issue on Smart Cities and the Future Internet in Europe. *Journal of the Knowledge Economy*, Vol. 4(2), 119–134.

Komninos, N., Panori, A., and Kakderi, C. 2021. The Smart City Ontology 2.0: Assessing the Components and Interdependencies of City Smartness.

Korczak, J., and Kijewska, K. 2019. Smart Logistics in the Development of Smart Cities. *Transportation Research Procedia*, Vol. 39, 201–211.

Kumar, A., Hooda, S., Gill, R., Ahlawat, D., Srivastva, D. and Kumar, R. 2023. Stock Price Prediction Using Machine Learning. In *2023 International Conference on Computational Intelligence and Sustainable Engineering Solutions (CISES)*, pp. 926–932. IEEE.

Lupiañez Villanueva, F. 2017. *CiudadesInteligentes: Evaluación Social de Proyectos de Smart Cities*. Montevideo, Uruguay Centro de Estudios de Telecomunicaciones de América Latina. https://cet.la/estudios/cet.la/ciudades-inteligentes-evaluacion-social-proyectos-smart-cities/

Ma, Shaojun, Li, Lei, Ke, Huimin, and Yilin, Zheng. 2022. Environmental Protection, Industrial Structure and Urbanization: Spatiotemporal Evidence from Beijing–Tianjin–Hebei. *China Sustainability*, Vol. 14(2), 795. https://doi.org/10.3390/su14020795

Martínez, I.P., and Alvarez, R.M. 2019. Importance of Research Ethics Committees in Family Medicine. *Prima*, Vol. 51, 263–265.

Mensah, J. 2019. Sustainable Development: Meaning, History, Principles, Pillars, and Implications for Human Action: Literature Review. *Cogent Social Sciences*, Vol. 5, 1653531.

Miola, A., and Schiltz, F. 2019. Measuring Sustainable Development Goals Performance: How to Monitor Policy Action in the 2030 Agenda Implementation? *Ecological Economics*, Vol. 164, 106373.

Nam, T., and Pardo, T. A. 2011a. Conceptualizing smart city with dimensions of technology, people, and institutions. *Proceedings of the 12th Annual International Digital Government Research Conference: Digital Government Innovation in Challenging Times*. pp. 282–291.

Nam, T., and Pardo, T.A. 2011b. Smart City as Urban Innovation: Focusing on Management, Policy, and Context. pp. 185–194. Tallinn, Estonia. *Proceedings of the 5th International Conference on Theory and Practice of Electronic Governance*.

Neirotti, P., De Marco, A., Cagliano, A. C., Mangano, G., and Scorrano, F. 2014. Current Trends in SMART CITY INITIATIVES: Some Stylised Facts. *Cities*, Vol. 38, 25–36.

Nerini, F.F., Slob, A., Engström, R.E., and Trutnevyte, E. 2019. A Research and Innovation Agenda for Zero-Emission European Cities. *Sustainability*, Vol. 11, 1692.

Öberg, C., Graham, G. and Hennelly, P. 2017. Smart Cities: A Literature Review and Business Network Approach Discussion on the Management of Organisations. *IMP Journal*, Vol. 11, 468–484.

OECD. 2017. Measuring Distance to SDG Targets. An Assessment of Where OECD Countries Stand. https://www.oecd.org/sdd/OECD-Measuring-Distance-to-SDG-Targets.pdf

Patton, M.Q. 2005. *Qualitative Research In Encyclopedia of Statistics in Behavioral Science.* Chichester, UK. John Wiley & Sons, Ltd.

Piattini, Mario, Calero, Coral, and Ruiz, Francisco. 2006. *Ontologies for Software Engineering and Software Technology.*

Pisanelli, D.M., Gangemi, A., and Steve, G. 2002. Ontologies and Information Systems: The Marriage of the Century? *Proceedings of LYEE Workshop.*

Praharaj, S., Han, J.H., and Hawken, S. 2018. Towards The Right Model Of Smart City Governance In India. *International Journal of Sustainable Development and Planning*, 13, 171–186.

Prashar, Neetika, Hooda, Susheela, and Kumar, Raju. 2023. Cybersecurity and Evolutionary Data Engineering: Current Status of Challenges in Data Security: A Review. *International Conference on Cybersecurity in Emerging Digital Era*, pp. 3–13. https://doi.org/10.1007/978-981-99-5080-5_1

Pratt, A.C. 2016. Creative Cities: The Cultural Industries and the Creative Class. *GeografiskaAnnaler Series B-Human Geography*, Vol. 90, 107–117.

Purvis, B., Mao, Y., and Robinson, D. 2019. Three Pillars of Sustainability: In Search of Conceptual Origins. *Sustainability Science*, Vol. 14, 681–695.

Rana, N.P., Luthra, S., Mangla, S.K., Islam, R., Roderick, S., and Dwivedi, Y.K. 2019. Barriers to the Development of Smart Cities in Indian Context. *Information Systems Frontiers*, Vol. 21, 503–525.

Roosa, S. and Stephen, A. 2020. *Sustainable Development Handbook.* 2nd ed. New York, USA. River Publisher.

Sachs, J.D., Schmidt-Traub, G., Mazzucato, M., Messner, D., Nakicenovic, N. and Rockström, J. 2019. Six Transformations to Achieve the Sustainable Development Goals. *Nature Sustainability*, Vol. 2, 805–814.

Salgado, A.C. 2007. InvestigaciónCualitativa: Diseños, Evaluación Del Rigor Metodológico y Retos. *Liberabit*, Vol. 13, 71–78.

Saunders, T., and Baeck, P. 2015. Rethinking Smart Cities from the Ground Up. London: Nesta [Online] https://ofti.org/wpcontent/uploads/2015/06/rethinking_smart_cities_from_the_ground_up_2015.pdf

Schneider, F., Kläy, A., Zimmermann, A.B., Buser, T., Ingalls, M., and Messerli, P. 2019. How Can Science Support the 2030 Agenda for Sustainable Development? Four Tasks to Tackle the Normative Dimension of Sustainability. *Sustainability Science*, Vol. 14, 1593–1604.

Scott, A.J. 2008. Resurgent Metropolis: Economy, Society and Urbanization in an Interconnected World. *International Journal of Urban and Regional Research*, Vol. 32, 548–564.

Seunghwan, Myeong, Jaehyun, Park, Lee, MinHe. 2022. Research Models and Methodologies on the Smart City: A Systematic Literature Review. *Sustainability*. https://doi.org/10.3390/su14031687

Sev, A. 2011. A Comparative Analysis of Building Environmental Assessment Tools and Suggestions for Regional Adaptations. *Civil Engineering and Environmental Systems*, Vol. 28, 231–245.

Studer, R., Benjamins, V.R. and Fensel, D. 1998. Knowledge Engineering: Principles and Methods. *Data & Knowledge Engineering*, Vol. 25, pp. 161–198.

Sim, Sung-Han, Lee, Jong-Jae. 2021. Special Issue on Smart City and Smart Infrastructure. *Sensors.* https://doi.org/10.3390/S21217064. United 4 Smart Sustainable Cities, UNECE and ITU. 2015. [Online]. https://www.itu.int/en/ITU-T/ssc/united/Pages/default.aspx

Taamallah, A., Khemaja, M. and Faiz, S. 2019. The smart city of Tunisia. In Smart City Emergence (pp. 421–433). Elsevier.

Trindade, E.P., Hinnig, M.P.F., da Costa, E.M., Marques, J.S., Bastos, R.C., and Yigitcanlar, T. 2017. Sustainable Development of Smart Cities: A Systematic Review of Literature. *Journal of Open Innovation: Technology, Market, and Complexity*, Vol. 3, 11.

Urfalı, T. and Eymen, A. 2020. Determining smart intersections for smart city applications using multi-criteria decision-making techniques. The International Archives of the Photogrammetry, Remote Sensing and Spatial Information Sciences, Vol. 44, 405–412.

Vinuesa, R., Azizpour, H., Leite, I., Balaam, M.; Dignum, V., Domisch, S., Felländer, A., Langhans, S.D., Tegmark, M. and Nerini, Fuso. 2020. The Role of Artificial Intelligence in Achieving the Sustainable Development Goals. *Nature Communications*, Vol. 11, 233.

Wang, X., Chan, C. and Hamilton, H.. 2002. Design of knowledge-based systems with the ontology-domain-system approach. *Proceedings of SEKE*. pp. 233–236.

Washburn, D., Sindhu, U., Balaouras, S., Dines, R. A., Hayes, N. M. and Nelson, L. E. (2010). Helping CIOs understand smart city initiatives: Defining the smart city, its drivers, and the role of the CIO. Cambridge, MA. Forrester Research.

Wendling, L.A., Huovila, A., zu Castell-Rüdenhausen, M., Hukkalainen, M., Airaksinen, M. 2018. Benchmarking Nature-Based Solution and Smart City Assessment Schemes against the Sustainable Development Goal Indicator Framework. *Frontiers in Environmental Science*, Vol. 6, 69.

Yan, Y., Wang, C., Quan, Y., Wu, G. and Zhao, J. 2018. Urban Sustainable Development Efficiency towards the Balance between Nature and Human Well-Being: Connotation, Measurement, and Assessment. *Journal of Cleaner Production*, Vol. 178, 67–75.

Yigitcanlar, T., Kamruzzaman, M., Buys, L., Ioppolo, G., Sabatini-Marques, J., da Costa, E.M. and Yun, J.H.J. 2018. Understanding 'Smart Cities': Intertwining Development Drivers with Desired Outcomes in a Multidimensional Framework, *Cities*, Vol. 81, 145–160.

5 Analysis of the Techniques Used in the Development of Smart Cities and Urbanization Systems

Rashi Sahay and Sameeksha Khare
Manav Rachna International Institute of Research and
Studies, Faridabad, India

5.1 INTRODUCTION: BACKGROUND AND DRIVING FORCES

In this chapter we will look into spatial analysis and its working. Spatial analysis plays a crucial role in the development and management of smart cities. Smart cities leverage technology and data to enhance the quality of life for their residents, improve sustainability, and optimize resource allocation. Spatial analysis(Maia et al. 2019), which involves examining and interpreting data with a geographic or spatial component, is essential for achieving these goals. Here are some key aspects of spatial analysis in smart cities:

Urban Planning and Design: Spatial analysis helps urban planners and designers make informed decisions about land use, transportation networks and infrastructure development. It allows them to assess the impact of proposed changes on the city's layout and aesthetics.

Transportation Planning: Smart cities use spatial analysis to optimize transportation systems. This includes traffic flow analysis, route planning for public transportation and identifying areas prone to congestion. By analysing mobility patterns, cities can implement more efficient transportation solutions (Franch-Pardo, Ivan, et al. 2020).

Environmental Sustainability: Spatial analysis aids in monitoring and managing environmental factors such as air quality, water quality and green spaces. It helps cities identify pollution sources, plan green infrastructure and assess the effectiveness of environmental initiatives, (Nesbitt, Lorien, et al. 2019).

Emergency Response: Spatial analysis is crucial for emergency services. It helps responders locate incidents, plan evacuation routes and allocate resources efficiently during emergencies like natural disasters or accidents.

DOI: 10.1201/9781003467892-5

Public Safety: Crime mapping and hotspot analysis are important components of spatial analysis in smart cities. Law enforcement agencies use this data to identify high-crime areas and allocate resources accordingly.

Infrastructure Management: Predictive maintenance models can help extend the lifespan of infrastructure and reduce maintenance costs (Kuller, Martijn, et al. 2018).

Resource Allocation: Spatial analysis aids in optimizing resource allocation. For instance, it can help determine the best locations for public services like schools, hospitals, and fire stations based on population distribution and accessibility.

Real-time Data: Smart cities often integrate real-time data from various sources, such as IoT sensors and mobile apps. Spatial analysis allows this data to be processed, visualized, and used for real-time decision-making.

Citizen Engagement: Spatial analysis tools can be used to engage citizens in city planning and decision-making processes. Interactive maps and geospatial data can help citizens provide input and feedback on various urban initiatives.

Predictive Analytics: By analysing historical and real-time data, smart cities can make predictions about future trends and challenges(Lamba et al. 2021), such as population growth, traffic congestion, or energy consumption. This information is valuable for long-term planning. The data can be analysed by using regression models (Ward, M. D., & Gleditsch, K. S. 2018).

Accessibility and Inclusivity: Spatial analysis helps identify areas with limited access to essential services, ensuring that urban development is inclusive and meets the needs of all residents, including those with disabilities.

Data Visualization: Effective visualization of spatial data is crucial for decision-makers and the public to understand complex urban challenges and solutions. Geographic Information Systems (GIS) are often used to create maps and visual representations of data (Mitchell, A. Griffin, L. S. 2021).

Spatial analysis in smart cities is a multifaceted approach that uses geospatial data and tools to enhance urban planning, resource management, and the overall quality of life for residents. It plays a central role in optimizing city operations, promoting sustainability and ensuring that cities are more efficient, accessible and resilient (Ghilani, Charles D. 2017).

5.2 URBAN PLANNING AND DESIGN

Urban planning and design using spatial analysis involves the application of geospatial data and analytical tools to make informed decisions about the layout, development and management of urban areas. Spatial analysis helps urban planners and designers create more efficient, sustainable and liveable cities. Here are some key aspects of urban planning and design using spatial analysis:

Site Selection: Urban planners use spatial analysis to identify suitable locations for various urban facilities, such as schools, hospitals, parks and commercial centres. They consider factors like accessibility, proximity to existing infrastructure, and the needs of the community.

Land Use Planning: Spatial analysis is essential for zoning and land use planning. Planners can analyse current land use patterns and assess the potential impact of proposed zoning changes on the city's development.

Density and Population Distribution: Spatial analysis helps determine optimal population density in different areas of the city. Planners can analyse population distribution and density to ensure that resources are allocated appropriately.

Transportation Planning: Urban planners use spatial analysis to design efficient transportation networks. This includes analyzing traffic flow, identifying congestion hotspots and planning public transportation routes to minimize travel times and reduce congestion.

Environmental Impact Assessment: Spatial analysis allows for the assessment of the environmental impact of urban development projects. Planners can identify areas prone to flooding, evaluate the impact on natural habitats and plan for green infrastructure.

Walkability and Accessibility: Planners use spatial analysis to assess the walkability and accessibility of urban areas. They consider factors like sidewalks, crosswalks, public transit access and the proximity of amenities to residential areas.

Urban Green Spaces: Spatial analysis helps identify suitable locations for parks, green spaces and urban forests. Planners can ensure that these areas are distributed equitably across the city and provide recreational opportunities for residents.

Infrastructure Planning: Planners use spatial analysis to manage and plan infrastructure like roads, bridges, utilities and public services. Predictive maintenance models can help prioritize infrastructure projects and reduce maintenance costs.

Demographic Analysis: Spatial analysis allows planners to analyse demographic trends and population growth. This information is essential for long-term city planning and ensuring that services meet the needs of the population (Schabenberger & Gotway 2017).

Visualization and Public Engagement: Geographic Information Systems (GIS) and other spatial analysis tools enable the creation of interactive maps and visualizations. These tools can engage the public and stakeholders in the urban planning process by making complex data more accessible.

Sustainability and Resilience: Planners can use spatial analysis to promote sustainability by identifying opportunities for renewable energy, energy-efficient buildings and water conservation. Additionally, spatial analysis helps cities become more resilient to climate change and natural disasters.

Historical Preservation: Urban planners can use historical spatial data to preserve and protect historical landmarks and districts. This ensures that the city's heritage is maintained while accommodating modern development.

Traffic Management: Real-time traffic data and spatial analysis tools are used to manage traffic flow, optimize signal timings and improve transportation efficiency.

5.3 TRANSPORTATION PLANNING

Transportation planning using spatial analysis is a critical component of urban and regional planning. It involves the use of geospatial data, tools, and techniques to design, evaluate, and optimize transportation networks, infrastructure, and systems. Spatial analysis in transportation planning helps improve mobility, reduce congestion, enhance safety, and promote sustainable transportation options. Here are some key aspects of transportation planning using spatial analysis:

Network Design and Expansion: Planners use spatial analysis to design and expand transportation networks. They analyse existing traffic patterns, population distribution, and land use to determine the most efficient routes and modes of transportation.

Traffic Flow Analysis: Spatial analysis helps assess traffic flow and congestion. It allows planners to identify bottlenecks, congestion hotspots, and areas with high traffic volumes. This information is used to optimize signal timings, lane configurations, and road capacity.

Accessibility Analysis: Planners ensure that transportation options are accessible to all residents, especially those in underserved areas.

Transit Planning: Public transit agencies use spatial analysis to plan routes, schedules, and stop locations. They consider factors like population density, employment centres, and transit demand to provide efficient and convenient public transportation options.

Multi-Modal Transportation: Spatial analysis helps promote multi-modal transportation systems that integrate various modes of transportation, including buses, trains, trams, cycling and walking. Planners identify opportunities for seamless transfers between modes.

Safety Analysis: Transportation safety is a critical concern. Spatial analysis is used to identify high-accident locations, analyse crash data, and implement safety improvements such as road design changes, traffic calming measures, and improved signage.

Environmental Impact Assessment: Planners assess the environmental impact of transportation projects using spatial analysis. This includes analysing air quality, noise pollution, and carbon emissions to mitigate negative effects and promote sustainable transportation.

Park and Ride Facilities: Spatial analysis helps identify suitable locations for park and ride facilities, where commuters can park their vehicles and use public transit for the remainder of their journey.

Pedestrian and Bicycle Planning: Planners use spatial analysis to design pedestrian-friendly sidewalks, crosswalks and bike lanes. This promotes active transportation and reduces reliance on automobiles.

Demand Modelling: Demand forecasting models use spatial analysis to predict future transportation needs. These models consider population growth, employment trends, and travel behaviour to plan for future transportation infrastructure and services.

Public Engagement: Interactive maps and visualizations created through spatial analysis tools help engage the public and stakeholders in transportation planning processes. This input is valuable for decision-making and project prioritization.

Real-Time Traffic Management: Spatial analysis is used to manage traffic in real-time by monitoring and analysing data from sensors, cameras and GPS devices. Traffic management systems can adjust signals and provide real-time traffic information to drivers.

Corridor Analysis: Planners use corridor analysis to evaluate and improve transportation corridors. This includes assessing the effectiveness of transit corridors, highway corridors, and other transportation routes.

5.4   ENVIRONMENTAL SUSTAINABILITY

Environmental sustainability using spatial analysis involves utilizing geospatial data, tools, and techniques to better understand, monitor and manage environmental factors and resources. Spatial analysis contributes to informed decision-making, effective resource allocation, and the development of strategies to promote sustainability and protect natural ecosystems. Here are some key aspects of environmental sustainability using spatial analysis:

Resource Management: Spatial analysis helps monitor and manage natural resources such as water, forests and soil. It enables the assessment of resource availability, usage patterns and depletion rates.

Environmental Impact Assessment: Spatial analysis is used to evaluate the potential environmental impacts of development projects, land use changes and industrial activities. It aids in identifying ecologically sensitive areas and minimizing negative consequences.

Habitat Conservation: Conservation efforts benefit from spatial analysis by identifying critical habitats, migration corridors and biodiversity hotspots. This information informs the establishment of protected areas and wildlife conservation initiatives.

Land Use Planning: Sustainable land use planning relies on spatial analysis to designate areas for agriculture, urban development and conservation. It aims to balance development needs with the preservation of natural landscapes.

Climate Change Mitigation and Adaptation: Spatial analysis supports climate change efforts by assessing greenhouse gas emissions, identifying carbon sinks (e.g., forests), and planning for climate adaptation strategies like flood mapping and heat island analysis.

Water Resource Management: Spatial analysis is crucial for managing water resources, including surface water bodies, aquifers and watersheds. It helps track water quality, identify pollution sources and optimize water allocation.

Energy Efficiency: Spatial analysis contributes to energy efficiency efforts by identifying opportunities for renewable energy generation, optimizing energy distribution networks and assessing the impact of energy projects on the environment.

Air Quality Monitoring: Environmental agencies use spatial analysis to monitor air quality, identify pollution sources, and develop strategies to reduce air pollution in urban areas.

Waste Management: Spatial analysis assists in optimizing waste collection and disposal systems by identifying suitable landfill locations, assessing recycling programs, and monitoring waste generation patterns.

Ecosystem Services: Spatial analysis helps quantify the ecosystem services provided by natural environments, such as water purification, pollination and climate regulation. This information supports decision-making that prioritizes ecosystem protection.

Coastal and Marine Planning: Coastal and marine spatial analysis aids in managing coastal resources, protecting marine habitats, and planning for sustainable fisheries and aquaculture.

Natural Disaster Risk Assessment: Spatial analysis is essential for assessing the vulnerability of regions to natural disasters like hurricanes, floods and wildfires. It assists in emergency preparedness, evacuation planning, and post-disaster recovery.

Public Health: Environmental factors have a significant impact on public health. Spatial analysis can be used to study the relationship between environmental variables (e.g., pollution levels, green spaces) and health outcomes.

Conservation and Restoration: Spatial analysis guides conservation and restoration efforts by identifying areas that require protection or restoration and assessing the effectiveness of restoration projects.

Education and Outreach: Interactive maps and geospatial data visualization tools engage the public and stakeholders in environmental sustainability efforts, fostering awareness and support for conservation initiatives.

5.5 EMERGENCY RESPONSE

Emergency response using spatial analysis involves the application of geospatial data, technology, and analytical techniques to improve the effectiveness and efficiency of emergency management and response efforts. Spatial analysis plays a crucial role in planning for emergencies, coordinating response activities, and making informed decisions during crises. Here are some key aspects of emergency response using spatial analysis:

Resource Allocation: Spatial analysis helps emergency responders allocate resources effectively by identifying areas with the greatest need during a disaster. This includes locating emergency shelters, medical facilities, and supply distribution centres in optimal locations.

Risk Assessment: Spatial analysis assesses the vulnerability and risk associated with various hazards, such as floods, wildfires, earthquakes and hurricanes. This information informs disaster preparedness and mitigation strategies.

Evacuation Planning: Spatial analysis aids in evacuation planning by identifying evacuation routes, safe assembly areas, and potential bottlenecks in the event of an emergency. It helps optimize evacuation plans to minimize traffic congestion and ensure public safety.

Damage Assessment: After a disaster, spatial analysis is used to assess and quantify the extent of damage to infrastructure, buildings, and natural resources. This information guides response efforts and resource prioritization.

Hazard Mapping: Emergency management agencies create hazard maps that illustrate the areas at risk from various hazards. These maps are essential for public awareness, land use planning and emergency response planning.

Real-time Incident Mapping: Spatial analysis tools are used to create real-time incident maps that track the progression of a disaster, such as a wildfire or a flood. This information helps responders adapt their strategies as the situation evolves.

Search and Rescue Operations: Spatial analysis assists in search and rescue operations by optimizing search areas based on factors like last known locations, topography and weather conditions. It helps prioritize areas where survivors are most likely to be found.

Logistics and Supply Chain Management: Spatial analysis is vital for managing the logistics of emergency response, including the transportation and distribution of supplies, equipment, and personnel to affected areas.

Communication Infrastructure Planning: Spatial analysis helps plan and maintain communication infrastructure, ensuring that emergency responders have reliable communication networks during crises.

Public Alerts and Notifications: Spatial analysis is used to deliver targeted alerts and notifications to affected populations, taking into account their proximity to the disaster and the recommended actions to take.

Healthcare Facility Planning: During health emergencies (e.g., pandemics), spatial analysis helps identify suitable locations for testing centres, treatment facilities, and vaccination clinics. It also assesses healthcare facility capacity.

Wildfire Prediction and Monitoring: Spatial analysis is used to predict the spread of wildfires based on weather, terrain and vegetation data. It assists in monitoring fire perimeters and guiding firefighting efforts (Kumar et al. 2023).

Floodplain Mapping: Floodplain maps generated through spatial analysis help communities understand flood risks and make informed land use decisions to reduce flood-related vulnerabilities.

GIS-Based Decision Support: Geographic Information Systems (GIS) provide decision support tools that allow emergency managers to visualize and analyse data in realtime, facilitating informed decision-making.

Recovery and Resilience Planning: After a disaster, spatial analysis supports long-term recovery and resilience planning by assessing the rebuilding process, restoring critical infrastructure, and implementing measures to reduce future risks.

Infrastructure management: Infrastructure management using spatial analysis is a critical aspect of urban planning and development. Spatial analysis, often facilitated by Geographic Information Systems (GIS), helps public agencies and organizations efficiently plan, monitor, maintain and optimize infrastructure assets. Here are some key aspects of infrastructure management using spatial analysis.

Asset Inventory: Spatial analysis is used to create comprehensive inventories of infrastructure assets, including roads, bridges, utilities, public buildings and more. Each asset is geospatially referenced and catalogued for effective management.

Condition Assessment: Infrastructure assets are regularly inspected and assessed for their physical condition. Spatial analysis helps prioritize maintenance and repair based on the asset's age, condition and criticality.

Predictive Maintenance: Spatial analysis tools enable predictive maintenance by analysing historical data and using models to forecast when maintenance is likely to be required. This approach can extend asset lifespans and reduce long-term costs.

Asset Lifecycle Management: Infrastructure assets go through a lifecycle that includes planning, construction, maintenance and eventual replacement. Spatial analysis aids in tracking and managing the entire lifecycle, ensuring assets are replaced or upgraded when needed.

Resource Allocation: Spatial analysis helps allocate resources effectively. For example, it allows public works departments to prioritize road repairs or utility maintenance based on factors like traffic volume, usage patterns and asset condition.

Performance Monitoring: Infrastructure performance can be continuously monitored using spatial analysis. For example, it can track traffic congestion on roads, water quality in distribution systems, or energy usage in buildings.

Utility Network Management: Spatial analysis plays a vital role in managing utility networks like water, electricity and gas distribution systems. It helps identify leaks, optimize network designs, and respond to outages.

Spatial Planning: Spatial analysis is integral to urban and regional planning, guiding the placement of infrastructure to support community growth, reduce congestion and enhance liveability.

Emergency Response: During infrastructure emergencies, such as a water main break or a gas leak, spatial analysis helps first responders locate the problem area and assess the potential impact on nearby assets and communities.

Geospatial Data Integration: Spatial analysis integrates data from various sources, including satellite imagery, GPS, sensors, and remote sensing, to create a holistic view of infrastructure assets and their environment.

Land Use and Zoning Compliance: Spatial analysis ensures that infrastructure development is in compliance with land use regulations and zoning ordinances. It helps identify any discrepancies and supports decision-making for development proposals.

Environmental Impact Assessment: Infrastructure projects can have environmental impacts. Spatial analysis assesses these impacts, helping planners identify ecologically sensitive areas and propose mitigation measures.

Cost-Benefit Analysis: Spatial analysis supports cost-benefit analyses for infrastructure projects. It helps weigh the costs of development or maintenance against the expected benefits in terms of improved service delivery and quality of life.

Transportation Infrastructure: For transportation infrastructure, such as roads and bridges, spatial analysis assesses traffic patterns, identifies congestion points, and optimizes design and maintenance for safe and efficient transport.

Facility Siting and Expansion: Spatial analysis assists in determining optimal locations for public facilities, like schools, hospitals, and parks, and in planning expansions based on demographic trends and community needs.

5.6   RESOURCE ALLOCATION

Resource allocation using spatial analysis is a process that involves the use of geospatial data and analytical tools to distribute resources in an efficient and equitable manner across geographic areas. Spatial analysis plays a crucial role in optimizing resource allocation in various fields, including urban planning, emergency management, healthcare and environmental conservation. Here are some key aspects of resource allocation using spatial analysis:

Urban Planning and Development: Spatial analysis helps urban planners allocate resources like housing, transportation infrastructure, and public services to different neighbourhoods based on factors such as population density, socioeconomic status, and accessibility.

Emergency Response and Disaster Management: During emergencies, spatial analysis supports the allocation of resources, such as medical supplies, personnel, and equipment, to the areas most affected by disasters. It helps identify the locations where immediate assistance is required.

Healthcare Resource Allocation: In the healthcare sector, spatial analysis assists in allocating healthcare facilities, such as hospitals and clinics, to serve populations effectively. It considers factors like population health needs, demographics and geographic accessibility.

Education Services: Spatial analysis is used to determine the optimal locations for schools and educational facilities based on population distribution, transportation accessibility, and educational needs of communities.

Environmental Conservation: Conservation organizations use spatial analysis to allocate resources for the protection and restoration of natural habitats. This involves identifying priority areas for conservation based on biodiversity, ecosystem services, and threats to the environment.

Agriculture and Food Distribution: Spatial analysis supports the allocation of resources in agriculture by optimizing the placement of farms, warehouses and distribution centres. It considers factors like soil quality, climate and proximity to markets (Alonso, William, 2017).

Public Transportation: Spatial analysis assists in optimizing public transportation routes and schedules, ensuring that services are distributed efficiently to serve the commuting needs of different communities.

Infrastructure Development: Resource allocation for infrastructure projects, such as roads, utilities, and public facilities, relies on spatial analysis to determine the best locations for new development and maintenance activities.

Environmental Impact Assessment: Before resource allocation, spatial analysis is used to assess the environmental impact of projects, such as mining, construction and energy development. It helps identify ecologically sensitive areas and informs mitigation strategies.

Energy Distribution: Spatial analysis aids in the allocation of energy resources, optimizing the placement of power generation facilities and the distribution of electricity or other forms of energy to meet demand.

Hazard Mitigation: Resource allocation using spatial analysis includes the allocation of resources for hazard mitigation measures, such as flood protection, wildfire prevention, and earthquake preparedness. It identifies high-risk areas and directs resources accordingly.

Social Services and Welfare: Social service agencies use spatial analysis to allocate resources for welfare programs, housing assistance and food distribution to areas with the greatest need.

Transportation Infrastructure Maintenance: Spatial analysis helps prioritize maintenance and repair activities for transportation infrastructure, focusing resources on roads, bridges, and transit systems that require the most attention.

Economic Development: Spatial analysis supports economic development by guiding the allocation of resources to areas that have the potential for growth, job creation and investment.

Conservation and Sustainability: Environmental resource allocation involves distributing resources to promote conservation, sustainability and the responsible use of natural resources.

5.7 REAL-TIME DATA ANALYSIS

Real-time data analysis using spatial analysis involves the continuous collection, processing, and interpretation of geospatial data as it is generated in real time. This approach is invaluable in various fields, from transportation and emergency management to environmental monitoring and public safety (Gutiérrez, Javier, et al. 2017). Here are key aspects of using spatial analysis with real-time data:

Transportation and Traffic Management: Real-time data, including GPS tracking and traffic sensors, is used for monitoring traffic conditions, identifying congestion, and optimizing traffic signal timing. Spatial analysis helps create dynamic routing and navigation systems that guide drivers to the least congested routes.

Public Transit Operations: Transit agencies use real-time data to track the location and movement of buses, trains, and other transit vehicles. Spatial analysis helps improve scheduling, optimize routes, and enhance real-time passenger information (El-Assi, W., Salah Mahmoud, M. and Nurul Habib, K., et al. 2017).

Emergency Response: Real-time data is essential for emergency services. During disasters or accidents, spatial analysis helps responders locate incidents, plan evacuation routes, and allocate resources effectively. It assists in coordinating responses based on current conditions.

Weather and Environmental Monitoring: Real-time weather and environmental sensor data are analysed using spatial analysis to assess conditions such as air quality, flood levels and wildfire movement. This information informs public safety warnings and emergency preparedness.

Environmental Conservation and Ecology: Real-time data from remote sensors and GPS tracking devices help researchers monitor animal movements, track environmental changes, and assess the effectiveness of conservation efforts. This information guides ecological studies and conservation strategies.

Fleet Management: Organizations with fleets of vehicles, such as delivery services or public transportation, use real-time data to track the location, speed, and condition of vehicles. Spatial analysis optimizes routes and schedules while monitoring vehicle health.

Healthcare and Disease Tracking: Real-time data, including reports from healthcare facilities and sensor data, is used to track the spread of diseases and monitor public health issues. Spatial analysis assists in identifying disease clusters and assessing healthcare resource needs.

Utilities and Energy Management: Real-time data from smart grids and sensors is analysed to manage and optimize energy distribution, detect power outages, and monitor water and gas infrastructure. Spatial analysis aids in identifying issues and directing maintenance crews.

Supply Chain and Logistics: Real-time data in logistics and supply chain management is used for tracking shipments, monitoring inventory levels, and optimizing transportation routes. Spatial analysis assists in managing the movement of goods efficiently.

Agriculture and Precision Farming: Real-time data from sensors, drones, and satellites is used for precision farming, monitoring crop conditions, and optimizing irrigation and fertilizer application. Spatial analysis supports decision-making in agriculture.

Retail and Marketing: Real-time data from customer behaviour, mobile apps and point-of-sale systems is used to analyze foot traffic patterns, optimize store locations, and personalize marketing campaigns based on location data.

Crowdsourced Data: Real-time crowdsourced data from mobile apps, social media, and other sources is analysed for various purposes, including traffic updates, public safety reporting and event monitoring (da Silva, Antonio Samuel Alves, et al. 2019).

Law Enforcement: Real-time data is crucial for law enforcement agencies, as it helps track and manage incidents, dispatch officers, and monitor critical events. Spatial analysis supports situational awareness and decision-making.

Aircraft and Maritime Tracking: Real-time data from radar, AIS (Automatic Identification System), and satellite tracking is used to monitor the movement of aircraft and vessels. Spatial analysis assists in ensuring safe and efficient operations.

Smart Cities: Real-time data is a foundational element of smart city initiatives, facilitating data-driven decision-making in urban planning, traffic management, and public service provision.

One major realtime analysis factor is the population of a geographical area. In this chapter we will use VENSIM software to analyse the population of India for a span of 100 years.

We will take the following parameters:

Span: 100 years.
Population of India: 100000000.
Birth rate: 2.5%.
Death rate: 2%.

As shown below, the relation of parameters represented on Vensim are as follows:

Equations Used:
{Births= ((Population of India* Birth rate)/100)}.
{Deaths= ((Population of India* Death rate)/100)}.
Vensim provides a platform that represents the data in a graph and also makes an entity-relationship diagram.

5.8   DATA VISUALIZATION

Data visualization using spatial analysis involves the use of geospatial data and technology to create visual representations of information, patterns and relationships that have a spatial component. These visualizations help individuals and organizations better understand, analyse and communicate complex spatial data. Here are some key aspects of data visualization using spatial analysis (Hooda, S., et al. 2021).

Maps: Maps are one of the most common forms of data visualization in spatial analysis. They can represent a wide range of information, including demographic data, land use patterns, environmental conditions, transportation networks, and more. Geographic Information Systems (GIS) software is often used to create and customize maps for various purposes (Zhou, Chenghu, et al. 2020).

Heat Maps: Heat maps use colour gradients to represent the intensity or density of a phenomenon over a geographic area. For example, a heat map can depict the concentration of real estate prices in different neighbourhoods or the distribution of air pollution levels in a city.

Choropleth Maps: Choropleth maps use different colours or shading patterns to display variations in data over administrative or geographical regions. They are frequently used to depict population density, voting patterns and other regional data.

Proportional Symbol Maps: Proportional symbol maps use symbols (e.g., circles, squares) that vary in size to represent quantitative data, such as the number of crimes in different neighbourhoods, with larger symbols indicating higher values.

Flow Maps: Flow maps illustrate the movement of people, goods, or information from one location to another. They are often used in transportation planning, migration studies, and supply chain management to visualize the movement of goods.

3D Visualizations: 3D visualizations can add an additional dimension to spatial data by representing features and terrain in three dimensions. This can be useful for urban modelling, landscape design and geological analysis.

Time Series Maps: These maps display how spatial data changes over time. They can show the progression of a disease outbreak, urban development, or changes in land cover over the years.

Satellite Imagery and Remote Sensing: Data visualization techniques are used to process and display satellite imagery and remote sensing data. This is essential for monitoring environmental changes, land use and disaster assessment.

Interactive Dashboards: Interactive dashboards combine various spatial visualizations, enabling users to explore data interactively. Users can pan, zoom, and select specific data points to gain deeper insights.

Story Maps: Story maps are a type of interactive visualization that combines maps with narrative text, images, and multimedia elements to convey a story or message. They are commonly used for educational purposes and to engage the public in important spatial topics.

Geospatial Data Infographics: Infographics that incorporate spatial data provide a visual summary of information on a specific topic, such as the environmental impact of a project or the economic contributions of a region.

Location Analytics: Location analytics tools allow organizations to visualize business data, customer behaviour and market trends on maps. This helps in making location-based business decisions, such as site selection or sales territory optimization.

Spatial Data Web Apps: Web applications that incorporate spatial analysis and data visualization enable users to access, explore and analyze geographic information over the internet. These apps can serve various purposes, from real estate search to disaster response.

Augmented Reality (AR) and Virtual Reality (VR): AR and VR technologies can be used for spatial data visualization, offering immersive experiences for urban planning, navigation and architectural design.

Data Animation: Animations of spatial data can show changes and trends over time. This is particularly useful for visualizing dynamic processes, such as the spread of a disease or traffic patterns.

After a disaster, spatial analysis is used to plan and rebuild communities in a way that ensures accessibility and inclusivity for all residents, including those with disabilities. Spatial analysis supports the development of wayfinding systems that help individuals, including those with disabilities, navigate complex environments, such as airports, shopping malls and city centres.

REFERENCES

Alonso, W. (2017). *A theory of the urban land market.* pp. 149–157.

Da Silva, A.S.A., Stosic, B., Menezes, R.S.C. and Singh, V.P. (2019). "Comparison of interpolation methods for spatial distribution of monthly precipitation in the state of Pernambuco". *Brazil Journal of Hydrologic Engineering.* 24(3), p. 04018068.

El-Assi, W., Salah Mahmoud, M. and Nurul Habib, K. (2017). "Effects of built environment and weather on bike sharing demand: a station level analysis of commercial bike sharing in Toronto". *Transportation* 44, pp. 589–613.

Franch-Pardo, I., Napoletano, B.M., Rosete-Verges, F. and Billa, L. (2020). Spatial analysis and GIS in the study of COVID-19. A review. *Science of the Total Environment,* 739, p. 140033.

Ghilani, C.D. (2017). *Adjustment computations: Spatial data analysis.* John Wiley & Sons.

Gutiérrez, J., García-Palomares, J.C., Romanillos, G. and Salas-Olmedo, M.H. (2017). "The eruption of Airbnb in tourist cities: Comparing spatial patterns of hotels and peer-to-peer accommodation in Barcelona". *Tourism Management.* 62, pp. 278–291.

Hooda, S., Lamba, V., and Kaur, A. (2021). "AI and Soft Computing Techniques for Securing Cloud and Edge Computing: A Systematic Review". *5th International Conference on Information Systems and Computer Networks (ISCON),* pp. 1–5. IEEE.

Kuller, M., Bach, P.M., Ramirez-Lovering, D. and Deletic, A. (2018). What drives the location choice for water sensitive infrastructure in Melbourne, Australia? *Landscape and Urban Planning.* 175, pp. 92–101.

Kumar, A., Hooda, S., Gill, R., Ahlawat, D., Srivastva, D., and Kumar, R. (2023). "Stock Price Prediction Using Machine Learning." *International Conference on Computational Intelligence and Sustainable Engineering Solutions (CISES),* pp. 926–932. IEEE.

Lamba, V., Hooda, S., Solanki, V., Ahuja, R., Ahuja, S., and Kaur, A. (2021). "Comparative Analysis of Artificial Neural Networks for Predicting Nifty 50 value in The Indian Stock Market." *5th International Conference on Information Systems and Computer Networks (ISCON),* pp. 15. IEEE.

Maia, R., Gruson, H., Endler, J.A. and White, T.E. (2019). pavo 2: New tools for the spectral and spatial analysis of colour in R. *Methods in Ecology and Evolution* 10(7), pp. 1097–1107.

Mitchell, A. and Griffin, L.S. (2021). *ESRI guide to GIS analysis: Spatial Measurements and Statistics.* Vol. 171. pp. 147–190. ESRI Press.

Nesbitt, L., Meitner, M.J., Girling, C., Sheppard, S.R. and Lu, Y. (2019). Who has access to urban vegetation? A spatial analysis of distributional green equity in 10 US cities. *Landscape and Urban Planning,* 181, pp. 51–79.

Schabenberger, O. and Gotway, C.A. (2017). *Statistical methods for spatial data analysis.* CRC Press.

Ward, M.D. and Gleditsch, K.S. (2018). *Spatial regression models.* Vol. 155. Sage Publications.

Zhou, C., Su, F., Pei, T., Zhang, A., Du, Y., Luo, B., Cao, Z., Wang, J., Yuan, W., Zhu, Y. and Song, C. (2020). COVID-19: Challenges to GIS with big data. *Geography and Sustainability.* 1(1), pp. 77–87.

6 5G-Oriented Positioning Technology

Priyanka Gupta, Rupali Gill, Sukhraj Singh,
Shubham Sharma and Udaibir Singh Bhathal
Chitkara University, Punjab, India

6.1 INTRODUCTION TO MMWAVE

Millimetre waves are electromagnetic waves having wavelengths that span from 1mm to 10mm and have frequencies that range from 30GHz to 300GHz. This is a very high-frequency range, even higher than microwave signals, which is why mmWave is also known as Extremely High Frequency (EHF) radiation. Figure 6.1 shows a table featuring different categories of electromagnetic waves, including millimetre waves for the purpose of comparison.

mmWave offers unparalleled spectrum availability compared to cellular and wireless local area network (WLAN) microwave systems that operate at frequencies below 10GHz. This means that mmWave networks can support a much larger number of uses and devices simultaneously. Distinctively, the unlicensed spectrum at 60Hz provides 0 to 100 times* more spectrum than what is available for users of Wi-Fi and 4G cellular systems that operate at carrier frequencies below 6GHz. 60GHz radio has been repeatedly used as a reference to mmWaves because it is one of the most promising bands for mmWave communication, offering a very wide available bandwidth which is essential for supporting high data rates. Additionally, 60GHz regulations allow for higher transmit power than other mmWave bands, which helps to overcome the higher path loss at 60GHz. However, the high path loss at 60GHz also limits its range and makes it more susceptible to blockage, thus making it best suited for short-range applications such as indoor wireless local area networks (WLANs) and wireless personal area networks (WPANs).

Type of electromagnetic wave	Wavelength (meters)	Frequency (hertz)
Radio waves	10 meters to kilometers	30 kilohertz to 3 gigahertz
Microwaves	1 millimeter to 1 meter	300 gigahertz to 300 gigahertz
mmWave	0.001 meter to 0.01 meter	30 gigahertz to 300 gigahertz
Infrared waves	700 nanometers to 1 millimeter	430 terahertz to 300 terahertz

FIGURE 6.1 Different types of electromagnetic waves.

DOI: 10.1201/9781003467892-6

6.2 CHARACTERISTICS OF MMWAVE PROPAGATION

A crucial component of wireless communication is radio wave propagation, which affects a number of factors including attainable distances, interference levels, transmitter power, antenna specifications, and receiver design. Most physical things are quite large in relation to wavelength at mmWave frequencies, when wavelengths are less than a centimetre. Prominent propagation phenomena including reflection/scattering and signal blocking (shadowing) result from this. Additionally, because the wavelengths at mmWave frequencies are so narrow, the molecular makeup of water and air plays a significant influence in determining the maximum free space distances that may be reached across the 5G spectrum, with oxygen and water molecules causing dramatic attenuation of electromagnetic waves at specific frequencies. Besides these, humidity and temperature also affect these effects. It may be noted that identical to an inhibition of radio waves, mmWaves are essentially prone to human shadowing. They also possess restricted saturation through the majority of resources in correlation to waves at lower frequencies.

6.2.1 ATMOSPHERIC EFFECTS

Particularly at frequencies beyond 60 GHz, atmospheric attenuation is the dominant propagation restriction for mmWave signals. It is the signal that is absorbed by the radio atmosphere gas molecules like water vapour and oxygen. The atmospheric attenuation depends on the frequency, the status of the atmosphere as well as the length of the path. Oxygen absorption is the dominant factor in atmospheric attenuation at mmWave frequencies as shown in Figure 6.2. It has a peak at 60 GHz, so atmospheric attenuation is highest at this frequency. Water vapor absorption is also significant at mmWave frequencies, especially at frequencies above 100 GHz.

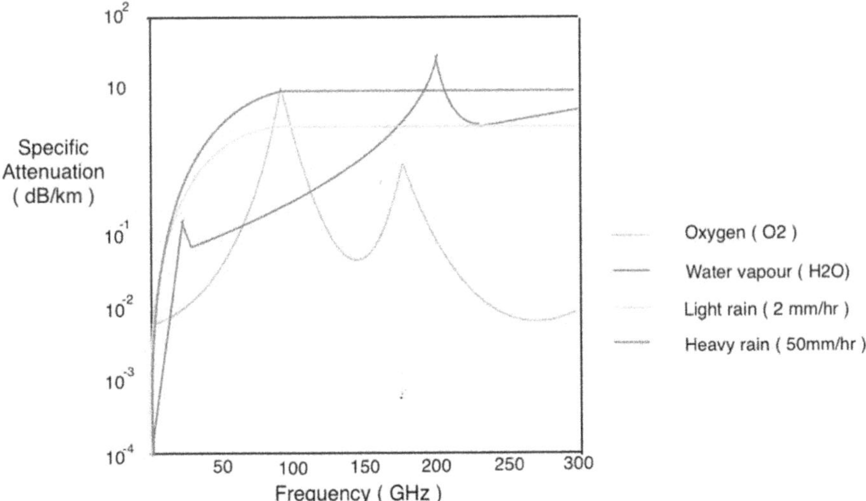

FIGURE 6.2 Specific attenuation curves of O_2, H_2O and rain as a function of frequency.

6.2.2 Beamforming for mmWave Communications

In the realm of millimetre-wave (mmWave) communications, operating within the frequency range of 30 GHz to 300 GHz, beamforming emerges as a pivotal solution to address the target user while nullifying interference, significantly improving the signal-to-noise ratio (SNR) at the receiver, thereby facilitating higher data transmission rates.

There exist three primary types of beamforming techniques: analogue, digital and hybrid. Analogue beamforming, the most straightforward and cost-effective, controls signal phases using phase shifters entirely in the analogue domain, though it is less flexible. Digital beamforming, executed in the digital domain using signal processors, offers high performance at a higher cost. Hybrid beamforming, a compromise between analogue and digital, utilizes analogue phase shifters for a subset of antennas and digital signal processors for the rest, providing a balance between performance and cost efficiency, making it particularly promising for mmWave Multiple Input Multiple Output (MIMO) systems.

6.3 MMWAVE POSITIONING

The emergence of fifth-generation (5G) mobile communication systems presents an opportunity to integrate mobile radio positioning, leveraging millimetre-wave (mmWave) technology to address the growing demand for wireless data and alleviate spectrum scarcity. Operating beyond 30 GHz, mmWave signals offer centimetre to millimetre-level location precision, with experimental 5G systems demonstrating promising results in urban environments. In parallel, 5G mmWave technology explores the potential of large bandwidths and antenna arrays for precise ranging and angle of arrival (AOA) and angle of departure (AOD) estimation, positioning 5G mmWave as the next-generation cellular positioning framework. This technology enables the creation of maps based on multipath component measurements related to physical objects. Cooperative positioning and mapping in 5G mmWave involve tasks such as vehicle positioning, environment mapping and information fusion among vehicles.

6.3.1 mmWave for Vehicular Positioning

The availability of different sensor technologies, communication protocols and advancements in mmWave bands are or will be adopted in the automotive domain, and this will ultimately lead to a smarter and more reliable transportation system. The complete understanding of the vehicle's environment and position is made possible due to the integrated application of technologies, such as lidar, cameras, radar and global navigation satellite-based systems. Mapping, a process of sensing and encoding of the environment by these onboard sensors, plays a huge role in diverse positioning applications, characterized by different requirements in terms of precision, latency, reliability and cost. While GNSS remains a cornerstone for vehicle absolute positioning as demonstrated in Figure 6.3, its limitations in urban canyons and beneath tree canopies compel the development of other technologies.

Various challenges are faced by Global Navigation Satellite Systems (GNSS) like barriers etc. effects in determining positions accurately. But with the use of mmWave technology elevated frequency carriers and highly developed antenna arrays facilitate

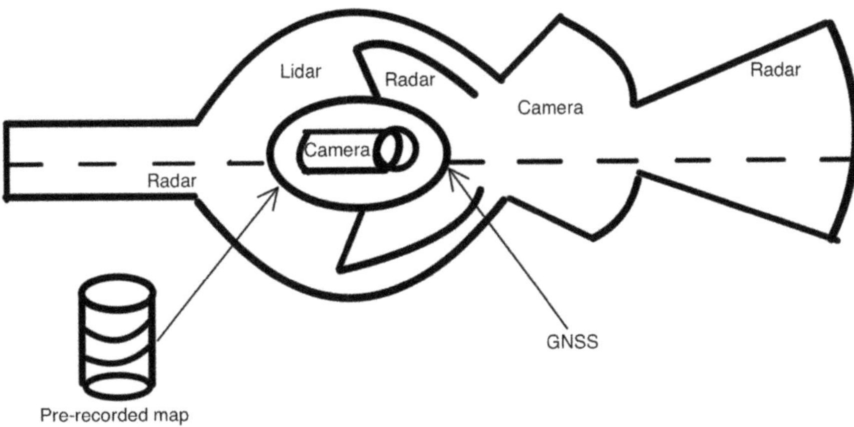

FIGURE 6.3 Combining GNSS and relative positioning for precise vehicle navigation.

to determine the precise positioning. Existing technologies like RADAR, GPS and LIDAR make a way to high precision positioning. Encounter difficulties in complex driving conditions. The appearance of V2X communications, e.g., V2I (Vehicle to Infrastructure), V2V (Vehicle to Vehicle) and V2P (Vehicle to Pedestrian) opens the door for driving services, based on V2X based positioning. This paper aims to provide an insight about the exploration of mmWave bands in V2X communication, from a communications and also from a positioning perspective. It identifies the challenges in exploiting mmWave for V2X communication and paves the way for further research and development.

6.3.2 Positioning Techniques

The ever-increasing development of the millimetre-wave (mmWave) systems has brought highly precise positioning and sensing to the threshold of the era. This is because of the fact that mmWave signals can produce perfect results regardless of path loss or environment providing accurate range and orientation estimations, thus being transformational in cooperative positioning and sensor calibration applications. The flexibility of mmWaves when worked together with sparse signal processing can really do a great job as it will lead to estimation of angle, delay and channel parameters altogether which results in higher location accuracy. As a result, in order to achieve improved localization accuracy, this technology is blended with GPS, cameras, and LiDAR systems. Hence, different vehicular applications combined together can provide both coverage and reliability for varying vehicle applications.

The positioning strategies in mmWave systems can be classified into: time-based, phase-based, power-based and angle-based methods. Power-based method use the path-loss model in the computations in order to determine separations, while in time-oriented approaches time-of-arrival (ToA) or time-difference-of-arrival (TDoA) are used to estimate the location. The phase-based technique such as Phase-difference-of-arrival (PDoA) employs phase variations for detecting the distance, while the angle-based approach like angle-of-arrival (AoA) and angle-of-departure (AoD) uses antenna arrays for decoding the propagation directions.

Challenges of positioning in mmWave communication systems involves evaluating the location of a user with respect to known reference points. Various methods, including ToA, TDoA, AoA and RSSI-based positioning, can be employed for this purpose. ToA-based methods require tight time synchronization between the transmitter and receiver, while AoA-based methods utilize directional received power to estimate the line of bearing. RSSI-based positioning utilizes the monotonic decrease of RSSI with distance to estimate range. By combining these methods, mmWave systems can achieve high-precision positioning with centimetre-level accuracy.

6.4 MASSIVE MIMO

6.4.1 TRADITIONAL MIMO

Traditional Massive IN – Massive Out was presented by Marconi in 1908. Provided by a few antennas on each end augmenting the network and data transmission volume for the end user. 3GPP LTE was based on the traditional MIMO technology, it was able to back up 8 single-stream users and 4 dual-stream users. Some of its limitations were restricted mobile terminal size, power depletion and physical aspect.

The evolution in wireless communication resulted in the increase of mobile users and to quench the increasing thirst of increased data rates of the end consumer the need to upgrade the Traditional MIMO came to light prompting up the Massive MIMO technology.

6.4.2 MASSIVE MIMO

Massive MIMO is structured in a way to use quite a few antennas at both transmitter and receiver end facilitating the parallel transfer of many data streams boosting the efficiency and reliability of network. Massive MIMO takes the traditional MIMO technology to the next level by using a massive number of antennas resulting in removing the problems introduced by the Traditional MIMO technology. In Figure 6.4, a basic Massive MIMO network is shown.

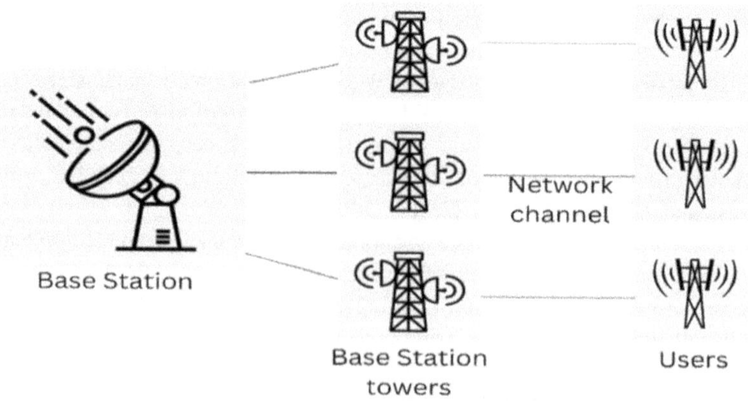

FIGURE 6.4   Basic massive MIMO network.

6.4.3 MASSIVE MIMO PRECODING TECHNIQUES

The major purpose of these precoding techniques is to augment the capability of the wireless network. These are used to interfere with factors such as spatial multiplexing, interference mitigation, beamforming, channel equalization, energy efficiency and data rate enhancement in the network.

There are 2 types of precoding techniques:

- Linear Precoding: It is used in MIMO wireless network systems to control the signals transferred from huge quantities of antennas to augment the system performance. It strives to pivot the signals to the specific user by eradicating the multi-user interference (MUI) for the specific and the other users in the networks as well.
- Zero Force Precoding Technology: ZF precoding is a LINEAR PRECODING technique is useful in the downlink network frameworks for the most part, i.e., the base station is loaded with great quantity of antennas and can impart quite a few data strings at the same time to numerous users. It is a type of spatial signal processing used in the MIMO wireless network for augmenting the system performance and terminating multi-user interface (MUI).

Zero Force strives to sketch a precoding system that transmits the signals to a specific user without any hindrance from the signals that are meant for the other users by transforming the imparted signal directly from the base station antennas.

Implementation of the Zero Force precoding technique is done by providing the BS the complete knowledge of the channel state information (CSI). Which is intricately captured and phase information between BS and each user. The information is attained through channel estimation techniques. At the instant after attaining the information of CSI, the precoding matrix, i.e., pseudo-inverse of the channel matrix, is calculated by the BS. Implementation of the new matrix system on the transferring of the new signals results in eradicating the multi-user interference. It can obtain perfect system capacity on two conditions, i.e., huge number of end users and ideal knowledge of the CSI. However, any fault in the calculation of CSI by the BS will result in a surplus hindrance and non-harmonical system performance.

The major advantages of the ZF are augmentation of the system performance, spectral efficiency and eradication of the multiuser interference. Although coming with the disadvantages such as its high responsiveness to the CSI error, high computational complexity and overhanging assessment for CSI estimation it is a dominant force in enhancing the system outputs in the MIMO wireless network with the consideration of its responsiveness to the CSI estimations in the process of hands-on implementations.

- Non-Linear Precoding: Non-Linear precoding technology strives to enhance the system performance by utilizing the non-linear integrity of the radio channel. Irrespective of the linear precoding technologies it applies non-linear functions to create and provide the signal-augmented malleability to enhance the system performance in specific scenarios.

- Signal-Level Precoding (SLP): Signal-Level Precoding is a nonlinear technique used to eradicate the signal disturbance and enhance the performance of system. To enhance the symbol-level performance SLP outlines the signals to make use of the cardinal symbol constellation.

In order to take advantage of the intrinsic interference present in the transmission network and alter the signals at the receivers SLP works toward fabricating the signal by taking into account the transfer of the data symbol and the channel proficiency.

Implementation of SLP is done in steps, the first step is the calculations of CSI by the Base station previously explained in ZF precoding technologies. The next is the inspection of the data symbols that are to be transferred between the different users. Based on scrutiny BS develops the precoding matrix specifically for each data symbol to modify the symbol level performance. In the final step the BS implies the matrices to the data symbol to manufacture the signals.

It can reach a higher production compared to linear precoding methods. It performs exceptionally well in the fields of higher interference. Its potential to mess with the interference results in the augmentation of symbol error probability (SEP).

Though this precoding technology comes with disadvantages such as restricted theoretical grasp, precise channel estimation and higher computational entanglement, its forte of messing up and making use of the network hindrance present in the system makes it a unique and favourable technique for the future MIMO system by keeping its higher computational complexity and accurate channel estimation into consideration.

6.4.4 Massive MIMO System Signal Detection Algorithm

The Massive MIMO network supplies its services to huge number of users. In the Massive MIMO system, the uplink signal can be detected when the signal is received by the base station from the terminal.

- Linear Detection: It is used when the base station is structured with a huge set of antennas. If the lower signal to interference plus noise ratio (SINR) is fulfilled the MRC receiver can carry out its functions without any issues but in case of large disturbance in network OLR will be optimized in comparison to the MMSE traditional receiver system.
- Nonlinear Detection: The detection algorithms based on Maximum Likelihood (ML) decoder are nonlinear detection algorithms such as Spherical Decoding (SD). The major drawback of these algorithms is they consider the nodes of a particular radius and have to augment the radius to detect any signalling point. Hence, adding only the most valuable nodes will decrease the search complexity.

6.5 DEVICE-TO-DEVICE COMMUNICATION

Device-to-Device (D2D) communication technology in the 5G network is an idea that has the power to completely transform the world (Farzamiyan, 2019). Some of its recognized applications include content distribution, relay communication,

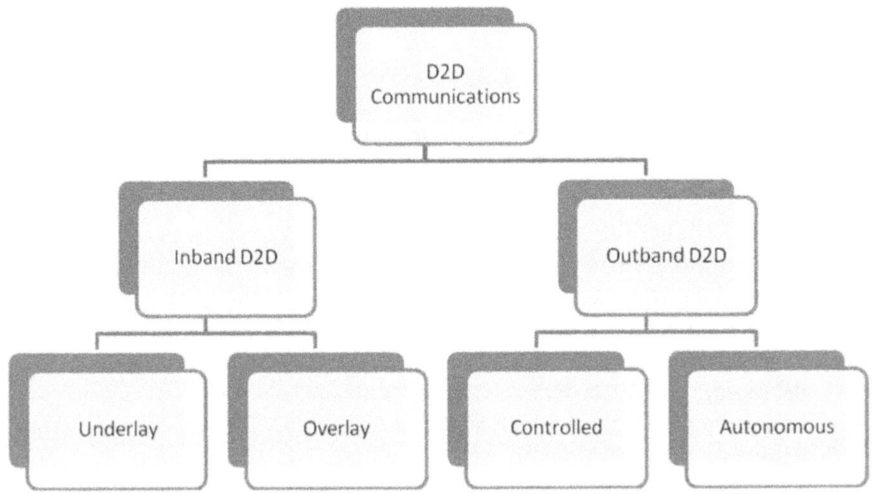

FIGURE 6.5 Device-to-device communication classifications.

cellular offloading, and machine-to-machine (M2M) networking (Ghavimi and Chen 2015). These will increase the coverage capabilities of base stations (BS), increase communication throughput and speed, decrease latency, and improve the energy economy of communication (Nwazor et al. 2022). 5G D2D can create and maintain a localized communication network without a base station. Three modes of operation are intended for the D2D communication architecture as shown in Figure 6.5 in the 5G network: in-band, out-band, and D2D relay communication (Hayat et al. 2019).

6.5.1 Device Discovery and Pairing Protocols

6.5.1.1 Device Discovery

Device Discovery – DD is a critical technique adopted by devices to identify and connect with neighbouring compatible devices (Fodor et al. 2014). According to Annanda Rathod, "Devices discovery – DD was used for natural environment observations, ocean observations, ecological observations." In addition to this, new and lesser-known applications including social networking, object tracking and finding, and mountaineer locating are being increasingly used in our daily lives (Ali et al. 2016a, 2016b). The recent applications of D2D – Device To Device time-bound include Proximity services, healthcare services and emergency services (Prashar et al. 2023).

6.5.1.2 Pairing

Pairing protocols are important for ensuring that we are creating reliable and secure connections between devices in D2D (Ali et al. 2018). D2D protocols are needed for identification, authentication and trust-building before devices are connected (Lianghai et al. 2017). Some of the prominent pairing techniques are Bluetooth Secure Simple Pairing (SSP), Wi-Fi Setup Pairing WPS, Certificate-Based pairing and Key Exchange protocols.

6.5.2 D2D COMMUNICATION FOR POSITIONING

6.5.2.1 Ranging Techniques

Ranging techniques are used in Device-to-Device (D2D) positioning to establish the relative positions or distances between devices (Hamoud et al. 2017). For D2D positioning, several ranging methods can be applied:

- Time of Flight (ToF): ToF calculates the time it takes for a signal to move between two devices (like radio signal or ultrasound). Devices measure the speed of the signal and the transit time to calculate the distance from one another. (Lin et al. 2014).
- Round-Trip Time (RTT): Like ToF, RTT measures the time taken for a signal to travel from one device to another and then back. One-way distance is half of round-trip time, which is effectively used to determine the distance between the devices involved. (Hunukumbure et al. 2013).
- Received Signal Strength Indicator (RSSI): RSSI measures the intensity of the received signal. With increasing distance between transmitter and receiver, the signal strength decreases. Thus, distance and strength are inversely proportional to each other (Militano et al. 2015).
- Angle of Arrival (AoA) and Angle of Departure (AoD): The RSSI measures the intensity of a received signal. It determines the signal strength of a wireless network. With increasing distance between transmitter and receiver, the signal strength decreases. Thus, distance and strength are inversely proportional to each other.
- Ultra-Wideband (UWB): UWB technology bests in measuring the time taken for signals to travel between transmission and reception. This is done by using brief pulses. Due to its exceptionally precise range measurements, UWB technology is suitable for indoor positioning applications (Alnoman and Anpalagan 2017).
- Beacon-Based Ranging: Devices can accurately identify their location by communicating with permanent nodes like anchors or beacons with known positions. Techniques, such as triangulation or trilateration, rely on exchanging signals between the device and the reference nodes to calculate the distance between them. This information is then used to determine the device's position in relation to reference points, providing accurate location data (Zhang et al. 2015).
- Differential ranging and Pseudorange ranging: These techniques calculate positions by using known reference locations or by measuring the time difference between signals sent from synchronized clocks (Chukhno et al. 2021).

6.6 ULTRA DENSE NETWORKS

The development of wireless networks within the present decade is because of the enlargement of cell phones and apps. The community design has reached a mileage with the operation of Ultra-Dense Networks. The UDN idea is an essential milestone on this street on the route of the community. Effectively, it's an important phrase which demonstrates the gravity of ultra-dense entry offering. The excessively dense community is established by the employment of small cells particularly speeded in condensed areas for the aim of connectivity. These small cells are poor in transference and protection. These are please distributed in lots of locations akin to avenue furniture, constructing or other extremely condensed locations. Enhanced spectrum effectuation and improved link high quality is a necessity to take care of the heightened community visitors.

Previously the model was trained with a 4G simulation that delivers 1000× cutting-edge WAN bandwidth. With a 5G network that is accompanied with gigantic MIMO and wave communication modelling. The first step is small cells positioning that could increase the network's extended throughput and impeccable coverage. The purpose of the UDN is spectrum handling, energy coordination and competition reflection and usage of resources.

6.6.1 Concept of Ultra-Dense Network

UDN depicts a deliberate modification in wireless network architecture focusing on controlling the conventional network restriction. They bring out-of-the-ordinary capacity, flawless coverage and ultra-low latency to the table. Featured by extraordinary compact network of small cells, much smaller and shorter ranged than the traditional cells. Concentrated cell deployment augments network capacity by structured and planned use of spectrum resource reuse. The packed deployment of the cells results in higher number of cell availability than the number of active users present in the area resulting in assignment of a whole single cell for augmented performance.

Sometimes it is also mentioned as USER CENTRIC ULTRA DENSE NETWORK structured in a way to learn and acclimate with respect to the user behaviour and movement pattern. The flexibility as mentioned above in this network provides the users QUALTY OF SERVICE (QoS) even in concentrated regions. With the given flexibility of resource distribution and modification of cell sizes UUDNs can provide a huge amount of user base while at the same time providing high reliability and harmonious user experience.

6.6.2 Features of Ultra-Dense Networks

A change to UDNs has been initiated from older networks with the positioning of small cells in dense areas in smaller range to users, i.e., higher number of smaller cells has been placed in small areas/regions near to the users for higher and reliable connectivity. This compelled the development in the network structure, management and positioning of the cells. Traditional frequency reuse techniques cannot be used in UDNs hence resulting in further development of spectrum operational techniques. UDNs are presumed of Line of Sight (LoS) transference because of lower range in base stations and users resulting in need for other broadcasting to precisely nullify the LoS components.

6.6.3 CHALLENGES

- High intrusion from other cells requiring structured mobility management also results in complexity of user association in dense network.
- The augmentation of an individual range of cell is also a challenge that remains unanswered.
- Increasing energy efficiency while maintaining high Quality of Service (QoS) and link quality gives rise to difficult modelling works, specifically in backhauling.
- As the number of cells increases in a network, this results in the augmentation of network entanglement.
- Instance information in context of authorization is transferred between users and AP groups, introducing difficulties in signal transference and affecting the network reliability and coverage.

6.7 CONCLUSION

It analyses the process and the favours of Massive MIMO technology. It also analyses the signal precoding and signal detection algorithms in brief. The Massive MIMO is a wireless network that is structured in a way to use a huge number of transmitters and antennas at the base stations for data transmissions. It offers many advantages over other technologies such as augmented data rates for users, increased network coverage, better signal quality, increased energy efficiency, reduced latency, easy spectrum handling and cheap rates.

Ultra-Dense Networks (UDNs) represent a paradigm shift in wireless networks, offering a promising approach to address the growing demand for wireless connectivity and support a wide range of applications. This chapter provides an overview of UDNs, highlighting their potential benefits and challenges. UDNs are characterized by a high concentration of small cells deployed in close proximity to users, enabling increased capacity, improved coverage, and reduced latency. They can be implemented in various scenarios, including HetNets, C-RANs, D2D networks, massive IoT environments, Massive MIMO networks, and mmWave networks. UDNs are expected to play a crucial role in the evolution of 5G and beyond, providing the necessary capacity and performance to support the ever-increasing demand for wireless data services.

The emergence of fifth-generation networks has rendered significant changes within the dimension of wireless communications. mmWave technology is a fundamental element that supports wireless communication. The paradigm shift that has been established within the 5G framework alludes to some of the unique challenges and opportunities in such contexts. The changes can be attributed to the proliferation of millimetre-wave frequencies within the 5G dimension. These waves are characterized by a smaller wavelength as well as a higher extent of carrier frequencies. Higher frequencies have also resulted in some wear and tear as they are passing through different probes. Such an aspect has significantly caused attenuation to the signal.

Despite the challenges, the potential benefits of mmWave technology are undeniable. By harnessing the vast spectrum and enabling high-speed data transmission, mmWave holds the power to revolutionize a wide spectrum of applications, encompassing immersive virtual reality experiences, autonomous vehicles and smart cities (Gill and Singh, 2020). As 5G networks continue to evolve, mmWave technology will undoubtedly assume an increasingly pivotal role in shaping the future of wireless communications. With ongoing research and innovation, mmWave holds the potential to transform the way we live, work and interact with the world around us.

Future Scope:
Massive MIMO is a befitting technology and ductile answer for the problems heading in way of wireless network standard. The features and the upper hand given by this in the field of networks quenches the thirst for higher data demands by the user, hence making it a befitted technology for 5G and downright for other networks above the 5G. Device-to-Device (D2D) communication has enormous promise for the future in a number of fields, with the following breakthroughs and developments possible:

- 5G and Beyond: More integration of D2D communication with 5G networks will allow for improved direct connections between devices.
- Edge computing and IoT: Direct-to-device (D2D) connectivity will play a key role in IoT ecosystems by enabling direct connection between IoT devices without exclusively depending on centralized cloud services (Hooda et al. 2021).
- Infrastructure and Smart Cities: D2D communication will be essential to the implementation of smart city projects, facilitating effective connection between sensors, infrastructure, and automobiles.
- Manufacturing and Industry 4.0: Machine-to-machine (M2M) connection will be made possible in industrial settings via D2D communication, which will result in more autonomous and effective manufacturing processes.
- Healthcare and Telemedicine: By enabling direct and secure connection between wearable sensors, medical equipment, and healthcare infrastructure, D2D communication will completely transform the healthcare industry. Reliable D2D connections will be beneficial for telemedicine, personalized healthcare services, and remote patient monitoring.
- Autonomous vehicles and vehicular networks: Vehicle-to-vehicle (V2V) and vehicle-to-everything (V2X) communication will be advanced by D2D communication, making transportation systems safer and more effective.
- AR and VR: Device-to-device (D2D) connectivity will improve AR/VR experiences by allowing cooperative and immersive interactions between adjacent devices. Direct connection between devices will be used in shared experiences, multiplayer gaming, and location-based augmented reality applications.

REFERENCES

Ali, K., Nguyen, H. X., Shah, P., and Bhuvanasundaram, N. 2016a. *"Architecture for Public Safety Network Using D2D Communication,"* IEEE Wireless Communications and Networking Conference, pp. 1–6.

Ali, K., Nguyen, H. X., Shah, P., Vien, Q. T., and Bhuvanasundaram, N. 2016b. *"Architecture for Public Safety Network Using D2D Communication,"* IEEE Wireless Communications and Networking Conference Workshops, WCNCW 2016, pp. 206–211.

Ali, K., Nguyen, H. X., Vien, Q. T., Shah, P., and Chu, Z. 2018. *"Disaster Management Using D2D Communication with Power Transfer and Clustering Techniques,"* IEEE Access 6, pp. 14643–14654.

Alnoman, A. and Anpalagan, A. 2017. *"On D2D Communications for Public Safety Applications,"* IEEE Canada International Humanitarian Technology Conference (IHTC), pp. 124–127.

Chukhno, Nadezhda, Oliver, Sergi, Torres-Sospedra, and Iera, Antonio. 2021. *"D2D-Based Cooperative Positioning Paradigm for Future Wireless Systems: A Survey,"* IEEE Sens. J. vol. 99, pp. 1.

Farzamiyan, Amir. 2019. *"A Survey on Device-to-Device Communication in 5G Wireless Networks." Porto. 14th Doctoral Symposium in Informatics Engineering.*

Fodor, G., Parkvall, S., Sorrentino, S., Wallentin, P., Lu, Q., and Brahmi, N. 2014. *"Device-to-Device Communications for National Security and Public Safety,"* IEEE Access, vol. 2, pp. 1510–1520.

Ghavimi, F. and Chen, H.-H. 2015. *"M2M Communications in 3GPP LTE/LTE-A Networks: Architectures, Service Requirements, Challenges, and Applications,"* IEEE Commun. Surv. Tutorials, vol. 17, no. 2, pp. 525–549.

Gill, R. and Singh, J. 2020. *"A Review of Neuromarketing Techniques and Emotion Analysis Classifiers for Visual-Emotion Mining,"* 9th International Conference System Modeling and Advancement in Research Trends (SMART), Moradabad, India, 2020, pp. 103–108, doi: 10.1109/SMART50582.2020.9337074.

Hamoud, Othmane, Kenaza, Tayeb, and Challal, Yacine. 2017. *"Security in Device-to-Device Communications (D2D): A Survey,"* IET Networks vol. 7, no. 1. doi:10.1049/iet-net.2017.0119.

Hayat, Omar, Ngah, R., and Zahedi, Yasser. 2019. *"In-Band Device to Device (D2D) Communication and Device Discovery: A Survey,"* Wireless Personal Commun. doi:10.1007/s11277-019-06173-9.

Hooda, Susheela, Lamba, Vikas, and Kaur, Amandeep. 2021. *"AI and Soft Computing Techniques for Securing Cloud and Edge Computing: A Systematic Review,"* Paper presented at 5th International Conference on Information Systems and Computer Networks (ISCON), Mathura, India, 2021, pp. 1–5. doi: 10.1109/ISCON52037.2021.9702422.

Hunukumbure, M., Moulsley, T., Oyawoye, A., Vadgama, S., and Wilson, M., 2013. *"D2D for Energy Efficient Communications in Disaster and Emergency Situations,"* 21st International Conference on Software, Telecommunications and Computer Networks, SoftCOM 2013, pp. 1–5.

Lianghai, J., Han, B., Liu, M., and Schotten, H. D., 2017. *"Applying Device-to-Device Communication to Enhance IoT Services,"* IEEE Commun. Stand. Mag., vol. 1, no. 2, pp. 85–91.

Lin, X., Andrews, J., Ghosh, A., and Ratasuk, R. 2014. *"An Overview of 3GPP Device-to-Device Proximity Services,"* IEEE Commun. Mag., vol. 52, no. 4, pp. 40–48.

Militano, L., Araniti, G., Condoluci, M., Farris, I., and Iera, A. 2015. *"Device-to-Device Communications for 5G Internet of Things,"* EAI Endorsed Trans. Internet Things, vol. 1, no. 1.

Nwazor, Nkolika and Ugah, Victory. 2022. *"Device-To-Device (D2D) Data Communications in 5G Networks." Int. J. Adv. Eng. Pure Sci.,* vol. 4, pp. 1151–1154.

Prashar, Neetika, Hooda, Susheela, and Kumar, Raju. 2023. *"Cybersecurity and Evolutionary Data Engineering: Current Status of Challenges in Data Security: A Review."* pp. 3–13. doi:10.1007/978-981-99-5080-5_1.

Zhang, Y., Pan, E., Song, L., Saad, W., and Han, Z. 2015. *"Social Network Aware Device-to-Device Communication in Wireless Networks,"* IEEE Trans. Wirel. Commun., vol. 14, no. 1, pp. 177–190.

7 Massive IoT for Industry Urbanization
Challenges, Emerging Trends and Use Case Scenario

Garima Chopra
Chitkara Institute of Engineering and Technology, Chitkara University, Rajpura, Punjab, India

Suhaib Ahmed
Model Institute of Engineering and Technology, Jammu, J&K, India

7.1 INTRODUCTION: BACKGROUND AND DRIVING FORCES

In the rapidly evolving realm of technology, the ascendancy of Internet-of-Things (IoT) applications has ushered in a new epoch of transformative advancements. These applications are characterized by their massive parallelism, data and computation intensity, and profound sensitivity to delays, span a diverse range, from autonomous driving to augmented and virtual reality online games, as well as the expansive domain of smart everything. As these applications progressively embed themselves into our daily lives, the imperative is for technology to advance at a pace that can meet their burgeoning demands.

However, the current stalwart in wireless communication, fifth-generation (5G) networks, though revolutionary, grapple with limitations in addressing the diverse needs of these ubiquitous IoT applications (Pons et al. 2023; Agiwal, Saxena, and Roy 2019). A noticeable gap exists between the capabilities of 5G networks and the prerequisites of these sophisticated applications, necessitating the emergence of sixth-generation (6G) networks. The promise of 6G lies in its high-level features, encompassing Terabits per second (Tbps) data rates, sub-millisecond (ms) latency, and centimetre-level localization, among others (Saad, Bennis, and Chen 2019; Chowdhury et al. 2020). These features are anticipated to bridge the existing chasm, facilitating the seamless operation of a vast network of IoT devices across a spectrum of service requirements. Drawing from an extensive survey of existing literature, this study identifies critical areas of concern in implementing IoT solutions at scale, delving into the challenges faced by both critical and massive IoT infrastructure and

DOI: 10.1201/9781003467892-7

discerning patterns and gaps in the current body of knowledge. Through this survey, key insights emerge, laying the groundwork for innovative approaches to address these challenges.

Against this backdrop, this paper embarks on a comprehensive exploration of massive IoT enabled by 6G networks, transcending the theoretical realm to delve deep into the existential challenges embedded within critical and massive IoT infrastructure. By scrutinizing existing literature, this study identifies not only challenges but also potential solutions, creating a bridge between theoretical understanding and practical implementation. Furthermore, the study illuminates emerging trends within the domain of 6G, outlining the trajectory of future developments. It is not merely a retrospective analysis but a forward-looking gaze into the technological horizon. The aim is not only to comprehend the current state of affairs but also to anticipate the future, ensuring that the proposed solutions remain relevant and effective in the face of evolving technologies. Through this real-world application, the study illustrates the transformative potential of 6G-enabled IoT in reshaping industries and societies. By delving into the practical application of 6G technology in the context of urbanization, the research elucidates how advanced IoT applications can seamlessly integrate into the fabric of our daily lives, revolutionizing industries and enhancing societal well-being.

The explosion of IoT devices has witnessed an unprecedented surge, driven by their ubiquitous integration into various industries such as transportation, agriculture and medicine, as documented in reputable sources (Zanella et al. 2014; Ahmed, De, and Hussain 2018; Da Xu, He, and Li 2014; Islam et al. 2015). In 2017, an estimated 8.4 billion IoT devices were in operation, marking a pivotal moment in the digital evolution of industries worldwide. This number, staggering in its own right, is merely a precursor to the impending IoT revolution. Projections indicate an imminent explosion in the IoT landscape, with a staggering 74.5 billion connected devices anticipated to be operational by the year 2025. This exponential growth can be attributed to the relentless march of technological advancement, coupled with the compelling need for enhanced efficiency, data-driven decision-making, and innovative solutions across diverse sectors. Industries have embraced IoT technologies as the linchpin of their operations, leveraging interconnected devices to optimize processes, reduce costs and enhance overall productivity. From smart transportation systems streamlining traffic flow to precision agriculture revolutionizing farming practices, IoT has permeated every facet of modern life.

The implications of this meteoric rise are profound. The interconnected web of devices promises a future where our environments are seamlessly integrated, facilitating a level of automation and intelligence previously unimaginable. Through the synergy of sensors, data analytics, and wireless connectivity, IoT devices are poised to revolutionize how we perceive and interact with the world around us. However, this monumental growth is not without its challenges. As the number of IoT devices skyrockets, concerns related to data security, privacy, and network infrastructure scalability become paramount. Addressing these issues necessitates innovative solutions, research, and collaborative efforts across academia, industry and policy-making bodies. The upcoming years will undoubtedly be defined by our ability to navigate these challenges and harness the full potential of IoT technologies for the betterment of society.

In essence, this research endeavours to bridge the gap between the growing requirements of advanced IoT applications and the capabilities of upcoming 6G networks. Through a meticulous analysis of challenges, trends and real-world applications, this study contributes significantly to the growing body of knowledge in the intersection of IoT and advanced wireless communications. By doing so, it not only advances academic understanding but also paves the way for a future where IoT applications operate seamlessly in a hyper-connected world, transforming the way we live, work and interact (Taneja et al. 2023).

7.2 EXISTING CHALLENGES IN INFRASTRUCTURE: CRITICAL IOT AND MASSIVE IOT

The term "critical IoT" describes the intentional incorporation of IoT technology in industries with the highest security, dependability and responsiveness standards (Schulz et al. 2017). This specific subset of IoT deployment includes applications vital to industries such as emergency services, transportation, energy and healthcare. Critical IoT is characterized by its capacity to operate flawlessly in high-stakes scenarios, frequently with redundant systems and fail-safe procedures guaranteeing continuous operation. These systems require quick decision-making, real-time data processing and minimal latency. Security is crucial, and there are strict safeguards in place to keep data integrity and guard against cyberattacks. Maintaining safety and dependability requires adherence to industry norms and laws. Accuracy, dependability and prompt reaction are prioritized in critical IoT deployments, which are essential for guaranteeing operational integrity and safety in vital sectors.

The implementation of critical IoT in industries requiring high dependability, security and real-time responsiveness is impacted by several obstacles. Security is still a top priority, requiring strong defences against illegal access and cyber threats. The requirement for redundant systems and fail-safe procedures makes it difficult to achieve dependability and resilience in these systems, especially in demanding or dynamic conditions. Achieving low latency and real-time responsiveness is difficult and demands quick data processing and device connectivity. Different IoT devices with non-standardized protocols provide interoperability problems that impede smooth integration. It is essential to ensure data integrity and quality, which calls for steps to verify data accuracy and eliminate corruption. Energy efficiency and power consumption management are ongoing challenges, while regulatory compliance and industry standards adherence add more complexity. In order to overcome these obstacles, coordinated efforts incorporating cutting-edge technology, strict security controls, standardized protocols, and continual research to improve the dependability and effectiveness of Critical IoT systems are needed. As seen in Figure 7.1, the comparison of various applications of critical and massive IoT is presented.

The healthcare industry, particularly hospital settings where patient monitoring and emergency response systems are critical, provides a strong use case for critical IoT. For example, in Hospital Emergency Response and Patient Monitoring, critical IoT is essential to guarantee patient safety, effective care delivery and quick emergency response in a hospital setting. Many IoT devices, including wearable sensors,

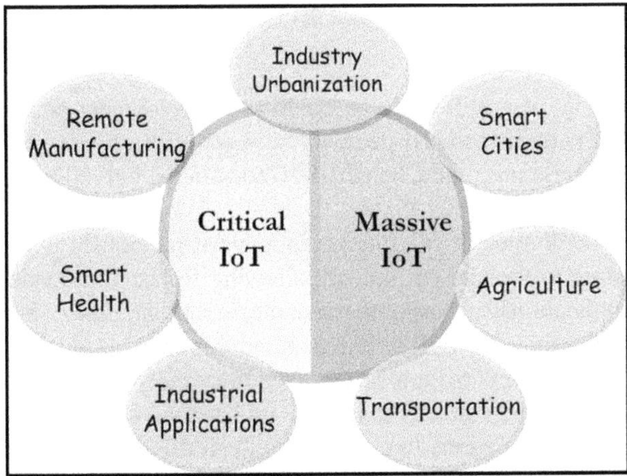

FIGURE 7.1 Comparison of critical IoT and massive IoT.

medical equipment and monitoring systems, are networked to monitor patients' health scenarios in real-time, evaluate patients' vital signs, and initiate prompt action as needed.

- **Patient Observation**: Patients' vital indicators, such as heart rate, blood pressure, oxygen saturation, and temperature, are continually monitored by wearable sensors. These sensors provide real-time updates to a centralized monitoring system by smoothly transmitting data through IoT-enabled devices.
- **Smart Medical Devices**: IoT-capable medical equipment with connectivity characteristics includes defibrillators, ventilators and infusion pumps. They provide remote monitoring and preventative maintenance to stop failures by sending operational data and status updates to a central system.
- **Emergency Response System**: An Internet of Things-driven emergency response system notifies medical personnel right away in emergency scenarios, such as abrupt changes in a patient's condition or alarms set off by medical devices. To ensure prompt action, the technology locates the patient and automatically notifies the appropriate medical personnel or equipment.
- **Predictive analytics** is made possible by large-scale patient data collection via IoT devices, which is combined with analytics and machine learning algorithms. These analytics assist in spotting possible patterns of health decline, allowing for preventive actions and early intervention to stop serious incidents.

By facilitating continuous monitoring, early health issue detection and quick emergency reaction, the integration of Critical IoT in hospitals greatly improves patient care. By creating a more secure and accommodating atmosphere for patient care, it streamlines healthcare processes, guarantees prompt interventions, and eventually enhances patient outcomes.

7.3 EMERGING TRENDS IN 6G: REQUIREMENTS AND FUTURE SCOPE

Most 6G technology was still in the research and development phase. Nonetheless, several projected patterns and prospective future applications were being considered (Figure 7.2) (Saad, Bennis, and Chen 2019; Chowdhury et al. 2020):

- **Extreme Data speeds**: 6G intends to achieve previously unheard-of data speeds of up to terabits per second, allowing for rapid uploads and downloads. Applications such as immersive media, holographic communications and high-definition AR/VR will depend on this speed.
- **Ultra-Low Latency**: 6G aims to provide practically instantaneous connectivity by focusing on latency in the sub-millisecond region. This is essential for real-time applications like remote surgery, driverless cars and sophisticated industrial automation.
- **Massive Connectivity**: 6G aims to support the Internet of Things (IoT) on a large scale by enabling connectivity for a massive number of devices per unit area. This might support billions of devices at once, enabling widespread connectivity, vast sensor networks and smart cities.
- **Energy Efficiency**: There is a lot of emphasis on increasing sustainability and energy efficiency. The goal of 6G is to provide technologies that can

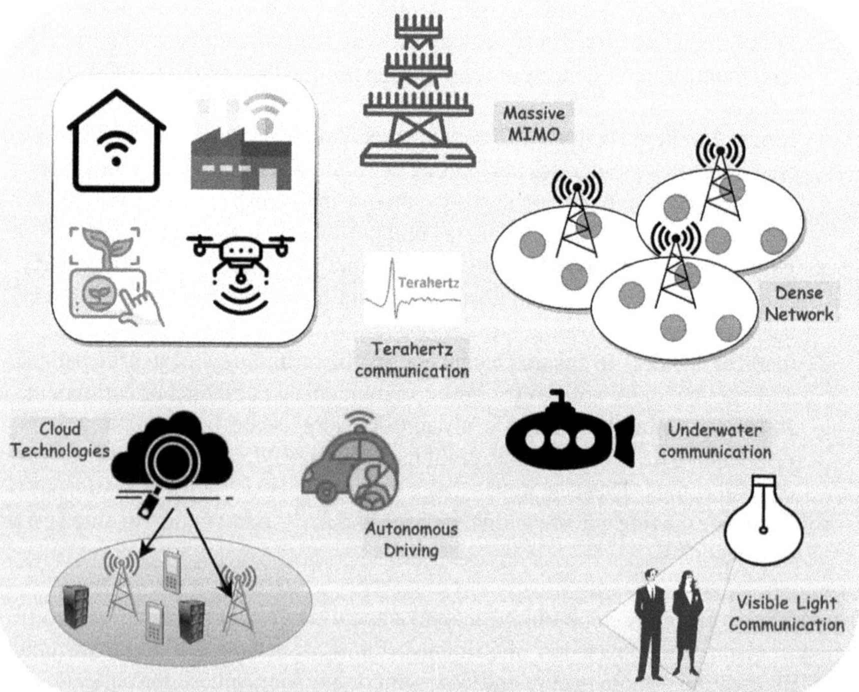

FIGURE 7.2   6G-enabled technologies.

manage the growing demand for connectivity and data while lowering the energy usage of wireless networks.

- **AI Integration**: It is projected that network management and optimization would incorporate artificial intelligence (AI). AI-powered resource allocation and network orchestration may improve user experiences, forecast traffic trends and improve network performance.
- **TeraHertz Frequencies**: Investigating the possibility of using THz frequencies for communication, which might greatly increase the amount of spectrum that is available and the capacity of data; nevertheless, there are difficulties associated with signal propagation and hardware design.

Beyond improved mobile broadband, 6G will eventually cover a wide range of sectors and industries. It seeks to transform not only communication but other industries such as manufacturing, transportation, healthcare and entertainment. Improved connectivity, incredibly low latency, and extensive device support may make it possible for ground-breaking uses including AI-driven technology improvements, highly immersive virtual worlds, seamless IoT integration, and remote surgery.

7.4 USE CASE SCENARIO FOR INDUSTRY URBANIZATION

This section overviews the use case scenario for Massive IoT by taking industrial application. Majorly, for industry, the massive IoT is used for "Asset tracking and Manufacturing facility."

For this case, the manufacturing facility is a cutting-edge production hub with a focus on producing intricate machinery and equipment. The facility is large and made up of several buildings, each of which has a distinct purpose in the production process. The processes involved in manufacturing are component development, assembly, quality assurance, and shipping. Every day, the plant processes a sizable amount of finished goods, work-in-progress components, and raw materials. The maintenance of optimal manufacturing processes, quality control and on-time delivery all depend on effective asset tracking and management.

7.4.1 APPLICATION OF MASSIVE IoT FOR INDUSTRY URBANIZATION

Massive IoT technologies have been adopted by facility management to address the difficulties of tracking and controlling assets in a large industrial facility. This can be accomplished by

7.4.1.1 Asset Tracking in Real Time

- IoT Tags and Sensors: All of the facility's assets have IoT tags or sensors installed. In addition to multiple sensors that record vital information like location, temperature, humidity and motion, these tags also have unique identifying numbers.
- Gateway Devices: The building is outfitted with strategically positioned IoT gateway devices to transmit the data from these tags to a central server. These gadgets act as the facility's communication centres.

7.4.1.2 Management of Inventory

- Centralized Database: A centralized database houses all the information pertaining to assets, such as components that are still being worked on, raw materials, and completed goods. Authorized personnel have access to this database from anywhere in the facility.
- Real-Time Updates: IoT sensors provide constant real-time updates on the locations and statuses of assets. This comprises details about whether an asset is being used in a particular manufacturing process, being stored, or travelling.

7.4.1.3 Automatic Alerting and Geofencing

- Digital geofences: Throughout the building, specific zones like assembly lines, QC areas, storage facilities and shipping docks are surrounded by digital geofences. These geofences specify the limits of certain procedures.
- Automated notifications: The IoT system sends out automated notifications whenever an asset crosses a geofence boundary. These notifications are forwarded to the appropriate staff members in charge of that particular region, guaranteeing prompt attention to the asset.

7.4.1.4 Health Monitoring of Assets

- Sensor Data Analysis: Temperature and humidity are two important environmental factors that the Internet of Things sensors on assets keep an eye on to guarantee the quality of sensitive components.
- Motion and Vibration Detection: In order to help detect any instances of misuse or mishaps, sensors also record information about movements and vibrations.
- Predictive Maintenance: To forecast the need for maintenance, the system examines sensor data. To lower the possibility of unplanned downtime, the system generates maintenance alerts when an asset shows wear or needs to be calibrated.

7.4.1.5 Reporting and Analytics

- Analysing Historical Data: The Internet of Things system gathers and examines historical data. Based on this study, patterns are found and process inefficiencies or bottlenecks are rectified.
- Custom Reports: With the data, authorized staff members can create bespoke reports. Inventory levels, maintenance schedules, transportation histories, and asset utilization are all included in these reports.

7.4.1.6 Connectivity with ERP

Smooth Integration: The facility's Enterprise Resource Planning (ERP) software and the IoT system interact with each other without any problems. Data interchange and end-to-end visibility between the two systems are made possible by this connection.

- Automated Workflows: Workflows can be made automated by integrating with ERP. For instance, the system initiates a restocking workflow when a certain component's inventory is low.

7.4.1.7 Mobile Programme
- Mobile Dashboard: A mobile application is accessible to staff members in various departments and responsibilities. A real-time dashboard displaying the locations and statuses of assets is provided by this application.
- Barcode Scanning: With their mobile devices, users can scan asset tags or barcodes. This tool makes it easier to check assets in and out of storage, find specific assets fast, and make sure they are handled properly.

7.4.2 ADVANTAGES OF MASSIVE IoT

Here is the list of the advantages by integrating Massive IoT with industry applications for urbanization:

- Enhanced Efficiency: By doing away with the need for manual searches, real-time asset tracking shortens lead times for production and streamlines the manufacturing process.
- Cost savings: By extending the lifespan of important assets and reducing maintenance costs, predictive maintenance and optimal asset utilization are achieved.
- Quality Control: Environmental monitoring keeps an eye on assets to make sure they are handled and kept properly, protecting delicate parts from harm.
- Precise Reporting: Adaptable reports offer practical advice for enhancing and streamlining procedures.
- Simplified Logistics: Timely deliveries are ensured by the effective routing of assets to various production phases, made possible by geofencing and real-time tracking.

This thorough scenario review demonstrates the difficulty in tracking and managing the assets in a contemporary manufacturing setting. Massive IoT is essential to ensuring that procedures run smoothly, cutting expenses, and producing high-quality goods on schedule. Enhanced productivity and operational efficiency are achieved through the smooth integration of real-time tracking, analytics, and predictive maintenance.

7.5 NEW EMERGING FIELDS FOR MASSIVE IOT

Massive IoT, a component of the wider IoT ecosystem, is being propelled by inventive applications and technology breakthroughs to keep growing and evolving into new and developing industries. The following are the new and emerging sectors where Massive IoT is having a big influence (Figure 7.3):

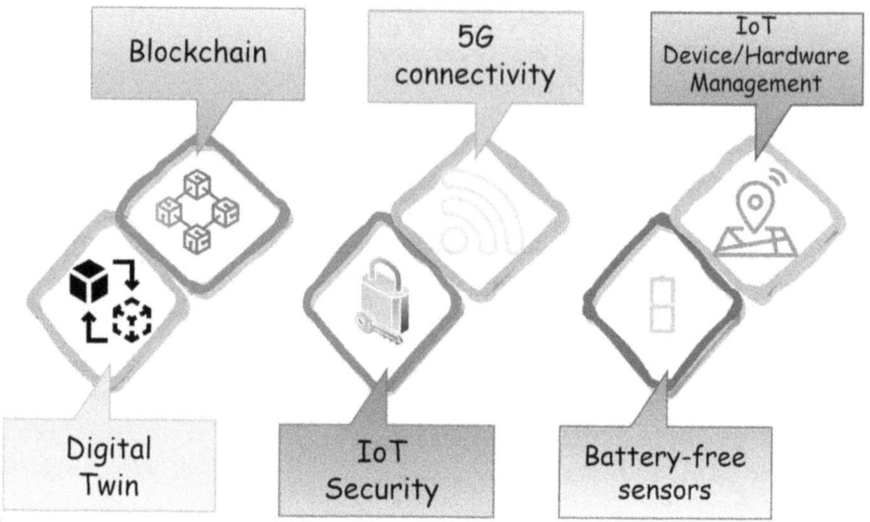

FIGURE 7.3　Emerging technologies in the field of IoT.

1. **Blockchain**: The combination of Blockchain technology and Massive IoT offers disruptive benefits in multiple domains. The integration of these technologies tackles important issues related to controlling, safeguarding and deriving value from the enormous amounts of data produced by networked IoT devices. This enormous inflow of IoT-generated data is guaranteed to be authentic and unchangeable by blockchain's decentralized ledger technology, creating a secure environment for data sharing among devices, systems, and stakeholders. In industries where provenance and accountability are crucial, such as supply chain management, this combination improves data security, transparency and traceability. Furthermore, the integration of smart contracts into Blockchain facilitates automatic, safe, and self-executing contracts or transactions between IoT devices, optimizing workflows and decreasing the requirement for middlemen (Singh et al. 2023). Even with its enormous potential, scalability, performance and energy efficiency issues still need to be addressed. As a result, research into how best to optimize Blockchain's handling of the massive amounts of data flow that characterize Massive IoT environments must continue. These technologies have the potential to completely transform industries as they develop and merge because they will transform data management, build trust and open up new possibilities for automation and efficiency in a variety of fields.

   Emphasis has been drawn to the use of blockchain in multiparty systems, particularly when coupled with the IoT (Ferrag et al. 2018). This combination addresses important privacy issues in smart grid transactions by providing safe storage and reliable sharing for IoT data. To manage the private calculation and aggregation of power data, a number of frameworks for power data transactions have been developed, utilizing blockchain as the transaction hub in conjunction with cryptographic algorithms (Liang et al. 2019; Gai et al. 2019). These frameworks are useful for gathering and analysing grid data, but they are not suitable for enabling the lossless transmission and storing of large amounts of IoT data.

Blockchain technology was successfully used by authors in Kang et al. (2018) to store and share data in the domain of Internet-connected cars. They used smart contracts and trust models to ensure that vehicles were exchanging high-quality data. Recent research emphasizes improving blockchain performance to meet IoT demands, rather than concentrating only on blockchain's role in IoT data collecting. To create an effective framework for IoT data exchange, for example, the authors in C. Xu et al. (2018) improved the blockchain consensus process and included new transaction kinds. But current frameworks often focus on particular features, such as performance improvement or data gathering, and don't offer a holistic view that takes into account all facets of IoT data management. A more comprehensive strategy is required, incorporating the security advantages of blockchain for sharing IoT data, resolving privacy issues, enhancing data transmission and customizing blockchain functionality to satisfy various IoT system needs.

2. **Digital Twin**: A virtual copy or representation of a real-world process, entity, system, or item is called a digital twin (Alnowaiser and Ahmed 2023; Liu et al. 2023). It is made to imitate and simulate the features, behaviour, and attributes of its physical counterpart utilizing real-time data from sensors, IoT devices, or other sources. With the use of this technology, it is possible to track, examine, and alter the virtual model in order to forecast, comprehend or improve the functioning of the actual system or object. Because they provide a comprehensive and dynamic representation of physical entities in a digital environment, digital twins find applications in a wide range of industries, supporting predictive maintenance, simulations, optimization and decision-making processes.

Digital twin technology has the potential to revolutionize many sectors by integrating with enormous IoT applications (Somers et al. 2023; H. Xu et al. 2023). This novel method entails building virtual representations of real-world IoT systems or devices, allowing for thorough study, simulation and optimization. Digital twins are essential in many industries, including manufacturing, healthcare, smart cities and agriculture. They provide resource optimization in agriculture, efficient urban planning, customized healthcare monitoring, and predictive maintenance in industrial installations. These twins also provide predictive analytics, which makes proactive decision-making possible in sectors like retail, telecommunications, and energy grids. Their ability to model real-world situations using IoT data enables organizations to anticipate problems, optimize workflows, and improve overall productivity significantly. Conclusively, digital twins hold immense potential in reshaping how we manage, monitor, and derive insights from extensive IoT ecosystems across diverse industries.

3. **Battery-free sensors**: The possibility of battery-free sensors to power Internet of Things devices without traditional batteries has drawn interest

(Bradai et al. 2020). These sensors don't require batteries to be changed or recharged because they capture energy from their surroundings. When combined with enormous IoT, they provide several benefits:

- *Durability*: By removing the need for throwaway batteries, cutting down on technological waste, and encouraging environmentally beneficial behaviour, battery-free sensors help to create more sustainable Internet of Things installations.
- *Low Maintenance*: When managing a large number of Internet of Things (IoT) devices, it is essential to minimize operational expenses and effort. Devices powered by energy harvesting techniques require little maintenance or replacement.
- *Scalability*: Battery-free sensors provide scalability in large-scale IoT deployments by removing the logistical burden of having to replace batteries in multiple devices dispersed over large areas.
- *Extended Lifespan*: These sensors can last longer because they don't require batteries, which age out with time and make IoT networks more dependable.
- *Flexibility in Deployment*: Since battery-free sensors are not constrained by power or battery life, they can be installed in a variety of settings, including isolated or difficult-to-reach areas.
- *Innovative Energy Harvesting Techniques*: These sensors provide for flexibility in design and deployment by drawing power from a variety of energy sources, such as solar, kinetic, thermal, or radio frequency energy.

Nevertheless, there are obstacles to overcome, including the restricted power availability from harvesting sources, the necessity for effective energy conversion and storage technologies, and lesser energy storage capacity in comparison to typical batteries.

However, battery-free sensors and large IoT have the potential to create low-maintenance, scalable, and sustainable networks of IoT throughout industries, revolutionizing data collection, transmission and utilization without relying on conventional power sources.

4. **IoT Security**: The deployment of IoT in daily life makes its future relevance clear. Its rapid growth is attributed to advancements in hardware approaches, such as the incorporation of cognitive radio-based networks to improve bandwidth and address underutilization of frequency spectrum (Wang, Uehara, and Sasaki 2015; Sicari et al. 2015). Wireless Sensor Networks (WSNs) and Machine-to-Machine (M2M) or Cyber-Physical Systems (CPS) have emerged in the literature as essential elements of the IoT. Because the IP protocol is the primary standard for connectivity, security issues pertaining to WSN, M2M or CPS persist in the context of IoT. Therefore, the entire deployment architecture must be protected against attacks that could compromise the IoT services or jeopardize data privacy, integrity or confidentiality. The IoT paradigm inherits the traditional security problems associated with computer networks since it is a collection of heterogeneous devices and interconnected networks. IoT security is further complicated by restricted resources because small devices and objects with

sensors have little power and memory. As such, the security solutions must be modified to fit the limited architectures (Khan and Salah 2018).

7.6 CONCLUSION

In this chapter, a review of massive IoT is conducted for industry urbanization. This chapter highlights the key aspects of massive IoT and critical IoT and also shows the comparison of both in terms of 6G applications. Then a discussion on 6G emerging applications and different challenges is carried out. A use case scenario for industry urbanization is also described in this chapter highlighting the key requirements with respect to massive IoT. It also highlights the emerging fields/applications for massive IoT and is discussed in detail.

REFERENCES

Agiwal, Mamta, Navrati Saxena, and Abhishek Roy. 2019. "Towards connected living: 5G enabled internet of things (IoT)." *IETE Technical Review* 36 (2): 190–202.

Ahmed, Nurzaman, Debashis De, and Iftekhar Hussain. 2018. "Internet of Things (IoT) for smart precision agriculture and farming in rural areas." *IEEE Internet of Things Journal* 5 (6): 4890–4899.

Alnowaiser, Kholood K, and Moataz A Ahmed. 2023. "Digital Twin: Current research trends and future directions." *Arabian Journal for Science and Engineering* 48 (2): 1075–1095.

Bradai, Sonia, Ghada Bouattour, Slim Naifar, and Olfa Kanoun. 2020. "Electromagnetic energy harvester for battery-free IOT solutions." *2020 IEEE 6th World Forum on Internet of Things (WF-IoT)*.

Chowdhury, Mostafa Zaman, Md Shahjalal, Shakil Ahmed, and Yeong Min Jang. 2020. "6G wireless communication systems: Applications, requirements, technologies, challenges, and research directions." *IEEE Open Journal of the Communications Society* 1: 957–975.

Ferrag, Mohamed Amine, Makhlouf Derdour, Mithun Mukherjee, Abdelouahid Derhab, Leandros Maglaras, and Helge Janicke. 2018. "Blockchain technologies for the internet of things: Research issues and challenges." *IEEE Internet of Things Journal* 6 (2): 2188–2204.

Gai, Keke, Yulu Wu, Liehuang Zhu, Lei Xu, and Yan Zhang. 2019. "Permissioned blockchain and edge computing empowered privacy-preserving smart grid networks." *IEEE Internet of Things Journal* 6 (5): 7992–8004.

Islam, SM Riazul, Daehan Kwak, MD Humaun Kabir, Mahmud Hossain, and Kyung-Sup Kwak. 2015. "The internet of things for health care: A comprehensive survey." *IEEE Access* 3: 678–708.

Kang, Jiawen, Rong Yu, Xumin Huang, Maoqiang Wu, Sabita Maharjan, Shengli Xie, and Yan Zhang. 2018. "Blockchain for secure and efficient data sharing in vehicular edge computing and networks." *IEEE Internet of Things Journal* 6 (3): 4660–4670.

Khan, Minhaj Ahmad, and Khaled Salah. 2018. "IoT security: Review, blockchain solutions, and open challenges." *Future Generation Computer Systems* 82: 395–411.

Liang, Wei, Mingdong Tang, Jing Long, Xin Peng, Jianlong Xu, and Kuan-Ching Li. 2019. "A secure fabric blockchain-based data transmission technique for industrial Internet-of-Things." *IEEE Transactions on Industrial Informatics* 15 (6): 3582–3592.

Liu, Xin, Du Jiang, Bo Tao, Feng Xiang, Guozhang Jiang, Ying Sun, Jianyi Kong, and Gongfa Li. 2023. "A systematic review of digital twin about physical entities, virtual models, twin data, and applications." *Advanced Engineering Informatics* 55: 101876.

Pons, Mario, Estuardo Valenzuela, Brandon Rodríguez, Juan Arturo Nolazco-Flores, and Carolina Del-Valle-Soto. 2023. "Utilization of 5G technologies in IoT applications: Current limitations by interference and network optimization difficulties—A review." *Sensors* 23 (8): 3876.

Saad, Walid, Mehdi Bennis, and Mingzhe Chen. 2019. "A vision of 6G wireless systems: Applications, trends, technologies, and open research problems." *IEEE Network* 34 (3): 134–142.

Schulz, Philipp, Maximilian Matthe, Henrik Klessig, Meryem Simsek, Gerhard Fettweis, Junaid Ansari, Shehzad Ali Ashraf, Bjoern Almeroth, Jens Voigt, and Ines Riedel. 2017. "Latency critical IoT applications in 5G: Perspective on the design of radio interface and network architecture." *IEEE Communications Magazine* 55 (2): 70–78.

Sicari, Sabrina, Alessandra Rizzardi, Luigi Alfredo Grieco, and Alberto Coen-Porisini. 2015. "Security, privacy and trust in internet of things: The road ahead." *Computer Networks* 76: 146–164.

Singh, Manmeet, Suhaib Ahmed, Sparsh Sharma, Saurabh Singh, and Byungun Yoon. 2023. "BSEMS—A blockchain-based smart energy measurement system." *Sensors* 23 (19): 8086.

Somers, Richard J, James A Douthwaite, David J Wagg, Neil Walkinshaw, and Robert M Hierons. 2023. "Digital-twin-based testing for cyber–physical systems: A systematic literature review." *Information and Software Technology* 156: 107145.

Taneja, Ashu, Shalli Rani, Jose Breñosa, Amr Tolba, and Seifedine Kadry. 2023. "An improved WiFi sensing based indoor navigation with reconfigurable intelligent surfaces for 6G-enabled IoT network and AI explainable use case." *Future Generation Computer Systems* 149: 294–303.

Wang, Yifan, Tetsutaro Uehara, and Ryoichi Sasaki. 2015. "Fog computing: Issues and challenges in security and forensics." *2015 IEEE 39th annual computer software and applications conference.*

Xu, Chenhan, Kun Wang, Peng Li, Song Guo, Jiangtao Luo, Baoliu Ye, and Minyi Guo. 2018. "Making big data open in edges: A resource-efficient blockchain-based approach." *IEEE Transactions on Parallel and Distributed Systems* 30 (4): 870–882.

Da Xu, Li, Wu He, and Shancang Li. 2014. "Internet of things in industries: A survey." *IEEE Transactions on Industrial Informatics* 10 (4): 2233–2243.

Xu, Hansong, Jun Wu, Qianqian Pan, Xinping Guan, and Mohsen Guizani. 2023. "A survey on digital twin for industrial Internet of Things: Applications, technologies and tools." *IEEE Communications Surveys & Tutorials.*

Zanella, Andrea, Nicola Bui, Angelo Castellani, Lorenzo Vangelista, and Michele Zorzi. 2014. "Internet of things for smart cities." *IEEE Internet of Things Journal* 1 (1): 22–32.

8 Iterative Approach to Implement and Deployment for Smart City and Urbanization

Monisha Gupta, Md. Sameeruddin Khan and Srinivas Mishra
Presidency University, Bangalore, India,

Preeti Sharma
Chitkara University, Punjab, India

8.1 INTRODUCTION: BACKGROUND AND DRIVING FORCES

The present cities continue to experience a number of societal issues, including environmental degradation and clogged roads. Additionally, the population of cities tends to increase over time, which will make those issues worse. A potential remedy for this is smart cities. As the name suggests, smart cities employ advanced communication techniques, sensors, and other electronic means to control their surroundings intelligently. This is where the Internet of Things comes into play. IoT is used to establish a strong connection between the networks, sensors, and devices that go into building a smart city. Unique Identifiers (UIDs) are present in every Internet of Things (IoT) system and are used to transmit data to any necessary network. One of the most promising and notable Internet of Things (IoT) applications is always thought to be smart cities (Okai et al. 2018) In current years, the perception of smart city plays an important field of interest in academia and industry, influencing the development and organization of various middleware phases. Urban communities need to become smarter and more connected in order to remain relevant. The largest urban communities in the world are being built at an astounding rate thanks to IoT (Kirimtat et al. 2020). Furthermore, "human to human" or "human to computer" communications won't be essential for system monitoring and control. Smart city uses Communication skills and information that are used to raise living standards while lowering expenses (Alberti et al. 2019). A growing number of people are turning to "smart" cities as a way to reduce the issues brought on by urbanization and population growth. The smart city concept, smart transportation, smart parks, smart homes, smart healthcare, etc. can all be designed many elements of smart city architecture are planned as one big network, allowing citizens to have access to data

DOI: 10.1201/9781003467892-8

management with varying levels of privacy. Smart cities provide infrastructure for collecting data from various sources, allowing for a better understanding about the city's state as well as the detection of problems and their potential solutions. They are a promising solution to these issues. The ability to foresee city status using knowledge gathered from both recent and historical data provides support of procedures which enhances management and overall standard of life in cities. IoT sensors and their interconnection have made significant strides in recent years (Al-Fuqaha et al. 2015), and many cities are already installed with equipment that may be used to gather data from several areas. Smart cities have their own set of difficulties, just like any other technology. The majority of connections created with RFID-based technologies are susceptible to hacking. Smart cities that are secure and resistant to hacking need to be improved. When smart cities are implemented, every one of our personal devices will be connected to a shared network, raising concerns about security and privacy because of hackers (Ahlgren et al. 2016). Furthermore, establishing smart cities comes at a hefty price. It is only through careful planning and effective equipment use that we will reap the benefits rather than the drawbacks. An increase in the potential use of machines puts simple, outdated jobs at risk. Because smart cities are being implemented, there is a possibility that unemployment will rise. The data needs to be accessible via multiple interfaces so that it may be accessed by diverse entities and applications. To assure that, a platform capable of handling a large volume of heterogeneous data (Arora et al. 2016), perhaps comprising sensitive information, is needed between the source of data and end applications. This platform must be able to manage scenarios with a high workload while retaining efficient and secure data organization and access (Londhe Rao 2017).

The "Aveiro Tech City Living Labs (ATCLL)" Core Data Platform will be the subject of this chapter. This platform will mainly focus on-

1. Diverse Information Sources: ATCLL has access to a wide range of information sources. These sources cover various fields, including:
   - Mobility Data: This may include data related to transportation and movement within the city, which is crucial for urban planning and traffic management.
   - Environmental Data: Information about environmental factors, such as air quality, noise levels, or weather conditions, can be valuable for sustainability and public health initiatives.
   - Network Data: Network data likely pertains to information related to communication and connectivity infrastructure, which can be essential for optimizing digital services in the city. Figure 8.1 depicts the model structure for ATCLL.
2. Core Data Platform: The presentation will likely delve into the Core Data Platform itself. This platform serves as the central hub for collecting, storing, and analysing the data from these diverse sources. It could also discuss how this data is processed and made available for decision-making and research.

The proposed architecture outlined in this chapter is intended for a genuine open platform that allows third party to gather data and test their individual solutions on a

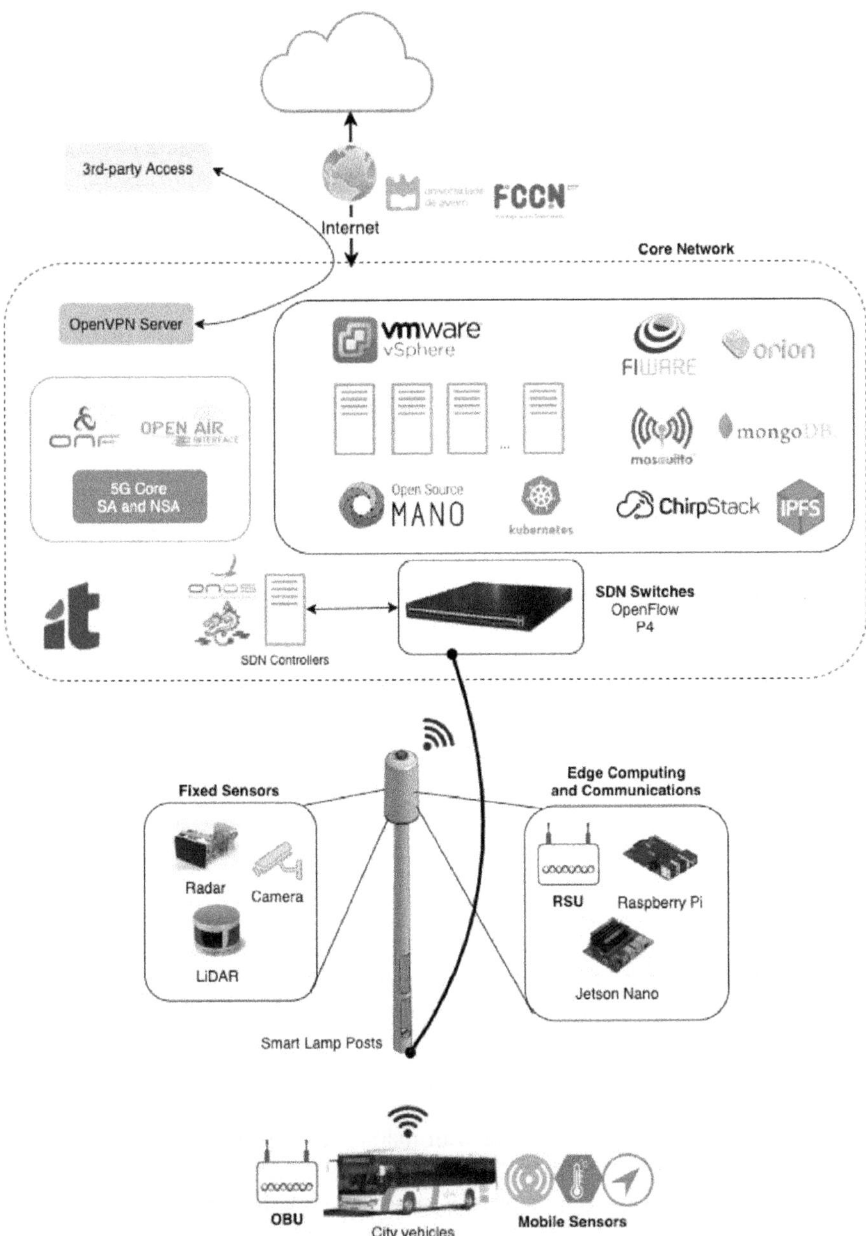

FIGURE 8.1 Model structure of ATCLL.

protected and accessible to exposed data policy (Vítor et al. 2021). We outline the data flow for data collection and retrieval while bearing in mind of IoT agents, access and authentication, data processing, real-time data transfer and diligence modules. Said data ranges from short frequency experimented data, for example, temperature and damp, to high frequency experimented data, for example moving vehicle location, heading and velocity, as determined via communication with moving automobiles, and sensor data in various areas throughout the cities (Zdraveski et al. 2017). The findings on the volume of data collected and data examples demonstrate how this policy can be utilized in the future to create new uses and add an intellectual layer for likelihood. This policy has been aimed to scale and remain successful as the volume of data increases. We updated the platform's design to allow for replication and distribution between physical and simulated machines in order to provide genuine redundancy and distribute the platform's overall load through the hosts. It was made likely through the use of cluster-based architecture. A cluster is a group of interconnected machines that work together to perform tasks. The use of a cluster-based approach likely enhances the platform's data processing capabilities and query performance. Mostly we are focusing on the scalability and persistency issues of the platform. The architecture previously used had all the services hosted on a single server, and Docker was used for deployment. This setup had some drawbacks, mainly single-point failure where hosting all services on a single host presented a significant risk of a single point of failure. If that host experienced issues or went down, the entire system could become inaccessible. The prior architecture made scaling the platform difficult because the volume of facts and the number of users grew over time. The passage emphasizes that the context in which this architecture is applied is a real smart city scenario. In such a scenario, there's a continuous increase in the amount of automobiles and flexibility sensors like video cameras and radars deployed in the city. This growth in data sources and users requires a more scalable and robust architecture. To address these challenges, a new architecture has been designed. It is built with horizontal scalability in mind. Key features of the new architecture include:

- Cluster-Based Approach- The new design employs a cluster-based approach. Clustering involves connecting multiple machines (hosts) to work together, distributing the load and ensuring redundancy. This architecture is inherently more fault-tolerant and scalable.
- Replication- Replication, which involves copying data and services across multiple machines, is used to enhance redundancy and ensure that the system can continue functioning even if one component fails.

8.2   PROPOSED DATA PLATFORM

Proposed model should be designed with mainly three goals: (i) creating a fibre-optic communication infrastructure that integrates a multi-protocol system with radio workstations; (ii) putting in place a recognizing platform that can comprehend how people behave in the city and offer innovative solutions for smart transportation systems, effective traffic supervision, and public safety, and lastly (iii) providing an exposed platform for other parties to validate their individual protocols, methods, prototypes and/or data analysis. The model structure is presented below.

1) **Access and control**:
   - Based on fibre equipment.
   - Edge computing capacity.
   - SDN network strategies with high data rates.

2) Sensing:
- Involves both static and mobile sensors.
- Mobile sensors may be present in vehicles.
- Collects various types of data, including mobility and environmental data.

3) Core and Backhaul:
- Comprises the data centres.
- Includes essential network strategies and servers.
- Responsible for handling the transportation of data from the edge to the central processing infrastructure.

4) Backend and Statistics Platform:
- Consisting of the core data platform.
- Involves a broker component for managing communication between different elements.
- Includes a database for storage of collected data.
- Incorporates data processing capabilities.

The proposed model network offers the fundamental infrastructure needed to capture all of the data that is collected by the fixed and mobile sensors and return it to the network's centre. In order to convey data to existent applications, preserving historic data, handling composed data for access by internal and external parties, and processing and analysing the data through projected and real-time data for usage by services and applications, an essential data policy is required (Montori et al. 2018). Since the data platform must be capable of receiving every kind of data while remaining active and open as it measures, the following questions and demands were addressed during its expansion:

(i) The most effective way to accumulate the various forms and kinds of facts that are received by the platform;

(ii) How to use a multi-tenant approach to maintain the information's organization by several domains;

(iii) How to make information available in real-time to services and apps via dynamic APIs;

(iv) How to maintain the security of the information that is available while maintaining the consistency of the data that enters the platform;

(v) How to mark the design of the architecture permanent and accessible to massive volumes of data;

(vi) How to make the platform accessible and efficient as the amount of sources grows and more facilities are built on best of it.

8.2.1   RELATED WORK

Smart cities have been the subject of numerous projects, most of which have concentrated on frame, data gathering and collection, or particular facilities. Here are a few instances of systems that provide information consistently. One such platform is CiDAP (Cheng et al. 2015), which was created to gather data after Smart Santander, some of Europe's major smart city test beds, besides providing it through a variety of interfaces to several applications. Another platform, 5G-SCSP (Kim et al. 2020), collects data from 5th generation-centred CCTV video and makes it accessible via several open APIs, enabling the creation of various services on the best of the policy. "IoTDA" is a combined Internet of Things data service that offers crucial data concerned with services (Lee et al. 2020), ranging from data collection to analysis based on deep learning. To get beyond the limitations of real-time processing, SOUL (Jung et al. 2017) proposes a stream reasoning system paradigm for the smart city application. This system was constructed using Apache Kafka, a message processing system, and Apache Storm, a real-time distributed processing system. MLK Smart Passage – a smart city test-bed that offers an open data policy (Harris et al. 2019) where scholars can contact generated statistics and an actual world testing situation for uses in areas including autonomous vehicles, pedestrian safety, and intelligent transportation. An examination of the ideal components of smart city policy and a demonstration of the "Sunrise Smart City Project," a practical implementation, are found in (Shahrour et al. 2017). The platform, which is housed at the "University of Lille's" scientific campus in "the north of France," gathers and processes the data from altogether over the campus and presents it in a UI that makes it possible to observe buildings and provides helpful information for university personnel.

The data platform utilized in the proposed model was inspired by these initiatives, which also helped identify the main areas on which it should be developed. Two important aspects of the data platform implementation that we emphasize are the integration of micro-services and the data's security and accessibility to third parties. The ATCLL platform integrates with an edge and cloud architecture to collect facts from mobile as well as static sensors together. It also tackles scalability and persistence, offering a fresh method for accessible data organization in large quantities of data scenario (Lee et al. 2019). Ultimately, our goal is to investigate a statistics platform result that can both support a range of communication policies and sensors by supporting multiple protocols and, on the one hand, offering a reliable and effective way to store and distribute the data collected from the proposed model infrastructure.

8.3 DATA PLATFORM

In order to address the various problems and specifications, we created the Core Data Platform. This platform has the ability to receive, store and process data from multiple domains, presenting it through a variety of secure and effective interfaces (Kolozali et al. 2019). This section will describe the platform's general architecture, how it easily satisfies the previously stated requirements, and highlight any flaws present in the platform's design.

8.3.1 ARCHITECTURE

The four primary modules of the proposed data platform are the IoT agents, accessing of data and its authentication, persistence of data, and real-time transport of data as depicted in Figure 8.2. Before exploring all the present modules in depth, it is crucial to know that platform was built on Orion Context Broker (Jung et al. 2017), serving as the core of the platform's only point of access and real-time transport

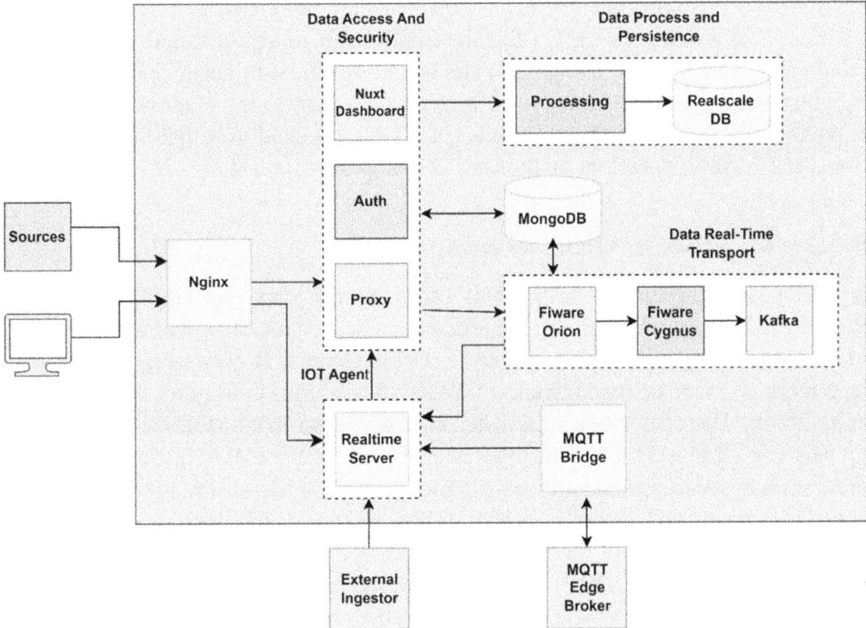

FIGURE 8.2   Representation of ATCLL data platform.

module. It follows publisher-subscriber architecture, allowing for the creation and updating of entities and subscriptions that are semantically separated by values included in the FIWARE-Service header of the requests, resulting in a multi-tenant model. This approach is extended in the proposed data platform, which additionally use the FIWARE-Service header for persistence, access and authentication, to information segregated by domains. The data platform also supports hierarchical scopes via FIWARE-Service Path, which can be utilized without restriction with the "Orion instance" to separate both entities and contributions of a "FIWARE" Service by diverse paths; however the remaining part of the platform, on the other hand, will not employ this unique header for different persistence/access or authentication. At the platform's entrance, a Nginx (Bellini et al. 2018) instance serves as a reverse proxy while handling SSL certificates and also allowing the servers to offload the decoding process.

8.3.2   INTERNET OF THINGS AGENT

According to Smart City context, there is a lot of variation in the data format between the various sources and tenants (Cheng et al. 2018). The platform requires one or more middleware services to intercept and convert data being received from bases that are not directly compatible with the appropriate data format (Harris et al. 2019). Only one IoT agent is being used, i.e., the "Real Time Server," which is Java-based server operating on spring boot, which is in charge of gathering data from various sources and converting it before sending it to the entry point. The "FIWARE" environment includes multiple IoT Agent that may be related to various interfaces such as "LoRaWAN" or "MQTT." These paths are mostly avoided because APIs would not directly adapt to edge info and would involve significant changes to report it, which means that whenever a fresh source is added to the platform, which isn't compatible with the NGSI model, wishes to be associated to actual Server or additional IoT agents dependent on the source's supported procedure.

8.3.3 REAL-TIME TRANSPORTATION DATA

A key component and single point of entry for new data into the platform is the Real-Time Transport, which is in charge of sending data to any platform subscriber in real time. A highly flexible broker i.e. Orion Context Broker which can manage high workload scenarios and link to FIWARE plugins plus third-party services (Badii et al. 2020). The agent uses a multi-tenancy paradigm that enables a rational split-up between all objects and contributions received via the FIWARE Service heading present on every requirement that we prolong across our whole platform to validate and persevere towards tenant information. Beyond this agent, there is a Kafka occurrence, likewise capable of high throughput that is connected to Orion via FIWARE-Cygnus (Shahrour et al. 2017) and spreads all the information that is being received at the open agent into remote network. Because this is the link to the outside, all facilities within the grid should practice this remote agent to reduce the burden on the open one. There is an MQTT Connection, a lower-resource MQTT broker that performs as a link to the edge computing statistics from the processed "Cooperative Awareness Messages (CAMs)," "Decentralized Environmental Notification Messages (DENMs)" and "vehicular standard messages" to the production of radar recognition, and distributes the acknowledged data to the Internet-of-Things Agent to reach the communal broker and further which will be made accessible to the external. Lastly, there is a Peripheral Ingestor, which is an entrance point for statistics from sources external to Aveiro (Wickramasekara et al. 2020).

8.3.4 ACCESS AND VALIDATION OF DATA

This component is in charge of maintaining the platform's authentication for the information entering and leaving, as well as providing persistent data via a number of dynamic APIs that don't need to be changed when a fresh resident joins the scheme. The FI-WARE structure includes a number of common enablers which enable the development of safety and validation in the FI-WARE atmosphere as well as for additional micro facilities that require safeguarding. "Key-rock IDM," "Authzforce" and "Wilma PeP proxy" are examples of generic enablers. "The Key-rock IDM" is used for storing characteristics and delivering "OAuth2" tokens, whereas "Wilma" is used for interrupting requirements to "Orion" or any extra protected service and validating the user authorizations using the provided token (Zhang et al. 2020). "Authzforce" supports the setting up of more advanced forms of authentication through the use of "XACML" Rubrics-based Consents. Even while these general enablers are fully acknowledged and offer ready-to-use Docker images, they don't

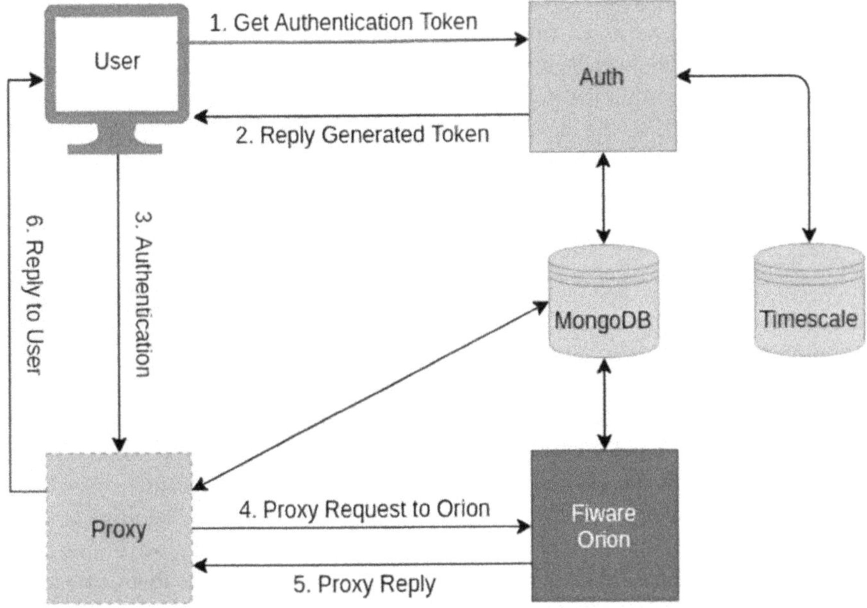

FIGURE 8.3 Data flow outside the data platform.

provide the mark of security that we require. As previously said, we are consuming Orion's multi-occupant model aimed at the entire policy, then these general enablers don't provide a direct mechanism to validate requirements for every tenancy, because "Authzforce" does not contact the request headers on its own. One way is to edit the main code of "Authzforce," which is preferable. As a result, we chose an alternative method, as illustrated in Figure 8.3. At the Validation "NodeJS" server also called Auth server, we created an account ideal with credentials like usernames and keys, with each version having authorization for one or more residents. These versions must be generated by a system administrator, who also manages the accounts that each occupant has access to. Following the creation of a version, the customer can produce a "JsonWebToken" via the "Auth" server, which will be legal for one time and that may be considered to mark any appeal to the "Orion" agent or other facilities that require authentication. The Orion and microservices are usually secured by the alternative server, which is also written in NodeJS and diverts all demands. In event of "Orion," it checks to see if the sign directed in the Permission title is authorized to access the occupier requested in the Fi-ware Service title. For further micro facilities, we applied the Alternative host in an active way, so that each call completed to that particular service needs a sign that includes an authorization to those occupants, that are not necessary to be directed at the "Fiware" Service heading of the requirements, that also provisions the same validation form as "Orion," verifying whether the sign has consent for the Fiware Service demanded. With these dual servers, we accomplished a verification architecture on uppermost of "Orion's multi-tenant architecture," in that we use Fiware Service not simply for the agent, including some other services that we need, while constructing it simple to add encoded facilities to the Substitute. This verification tool accomplishes the tenacity of defending the info and facilities within the policy, but it makes no promise that the info entering is legal. To report this, the supervisor can produce signs for each occupier, which must be utilized by the causes of the corresponding tenants when publishing data to "Orion," by including sending the produced sign at the Approval header, enabling us to retain the data entering policy reliably. This security system also protects the dashboard through the use of cookies, which allow users to verify themselves and obtain access to certain pages and facilities available on the dashboard (Prashar et al. 2023). In addition to the authentication option, this component is in charge of making the stored info available over various interfaces, notably active APIs. Those APIs were designed so that each tenant, whether new to the policy or not, can access them in a similar way, adjusting to further changes that may arise or their figures. These APIs are situated at the "Auth server," and these involve the Consent title to be directed in conjunction with the desired occupant, at the Fi-ware service title, to validate whether the consumer has authorization to request data after it. In addition to these secured API, the server publicly exposes certain information, which we utilize at the Console to demonstrate the various visualization and applications that can be built on the uppermost of the policy.

8.3.5 PROCESSING AND PERSEVERANCE OF DATA

In a typical Fi-ware design, the "Orion" context agent is linked to the Fi-ware Cygnus module, which is in charge of persisting information in the many storage services offered, such as MongoDB, CKAN, Apache Hadoop, and so on. This approach is considered as a way to persist the information, which requires a handful of subscriptions for the particular information that needs to be stored (Hooda et al. 2021). The solution stated above is not highly scalable for the long run. Cygnus components persevere information directly into the selected database without performing any settings on it, as a result, performance of query and usage of disk will worsen with time constraints. TimescaleDB differentiates the information collected from each tenant based on data throughput, and also provides data compression, reducing the usage of disk at a bare minimum level. Pre-aggregated tables continuously update the values and can be utilized for reducing the query execution time. Processing server creates databases and tables in response to incoming data, with configuration options for each tenant. Each tenant, specify data types, columns for persistence, and grouping options. Different batch sizes are allowed for each tenant's data types. This customization helps to optimize database operations by controlling the number of insert operations to avoid potential overhead. It is necessary for maintaining efficient database performance, especially as the amount of data and the number of tenants increases. TimescaleDB provides a setup of multiple nodes consisting of one access node and multiple data nodes. This setup is designed to scale well for accommodating the growth in the number of tenants, data volume and user interactions.

8.4   DATA FLOW DASHBOARD

This section describes how to use and manage the ATCLL Data Platform, including data flow graphs with historical data and console numbers. Figure 8.4 explains the entire process of data entering the platform. As previously stated, all information must be in NGSI format reaching the Orion. The sources that care and can deliver

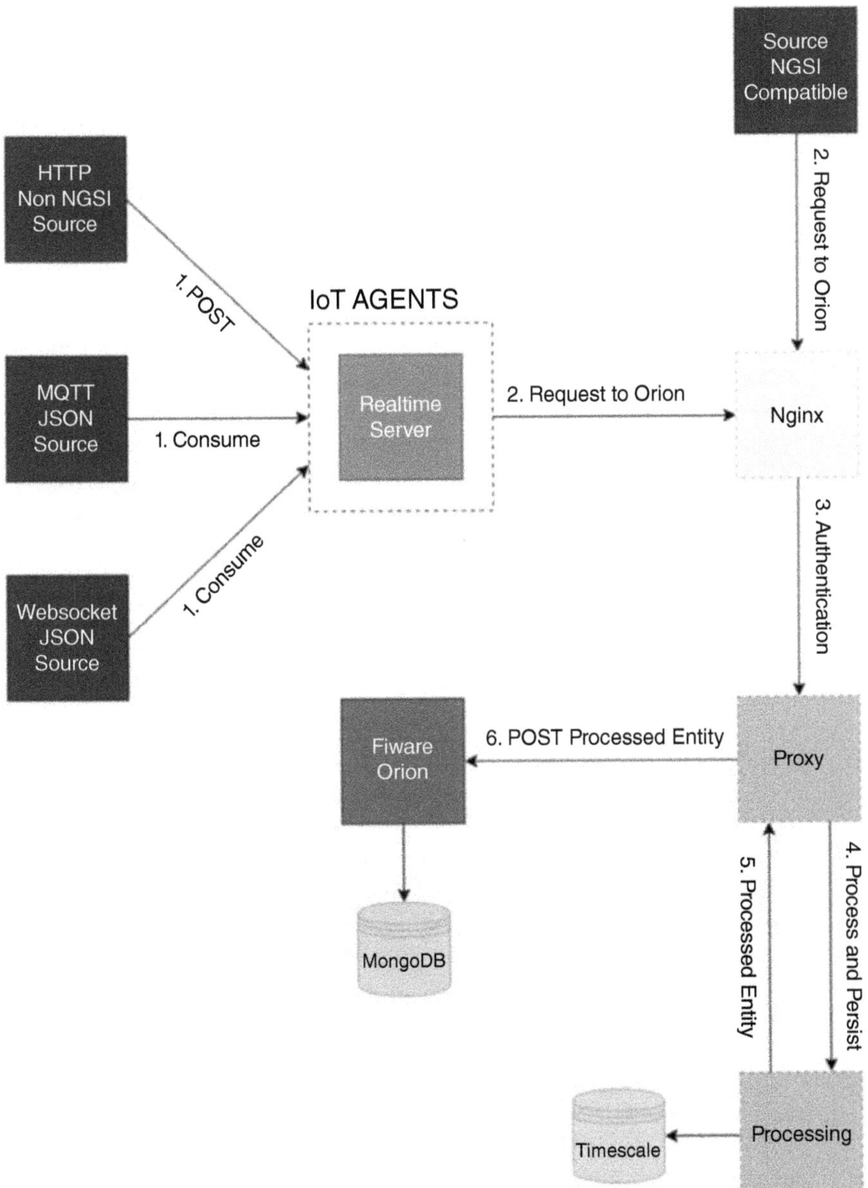

FIGURE 8.4   Data flow inside data platform.

data in "NGSI" organization are the simplest to instrument. Else, the IoT Instrument is required between the foundation and "Orion" to transform the format of the source to the NGSI format. In the form of a flow diagram, Figure 8.4 depicts how data enters the platform. Few of the objects displayed have rushed lines around them since they are not openly recognized by the foundations because they just send requests to the Orion broker and are unaware of the presence of any proxy server. If a fault occurs before any demand influences Orion, immediately the foundation will receive an HTTP answer indicating the disappointment or authentication issue. The data flow is described in the following steps:

1. The first step is considered as optional and is specifically designed for foundations that do not inherently support NGSI (Next Generation Service Interface) format directly or relevant for sources that need to use a transport method other than HTTP. The Real-time Server is responsible for consuming or receiving information from these non-NGSI sources and also capable of consuming data in any format from these sources. Later, it converts the received information into the NGSI format. Presently, the host consumes facts from a "Web Socket" host and an "MQTT" agent, both located within the network. WebSocket and MQTT are communication protocols, and the server is configured to interface with sources using these protocols. The system is designed to be adaptable, and it may be readily updated to accept HTTP requests from different sources if necessary. This adaptability is valuable in scenarios where new sources with different communication requirements may be introduced.

2. Data present in "NGSI" format, either at Real-time environment or at foundations that directly care about the plan, a request needs to be directed to Orion's support endpoint, which is interrupted by the "NGINX" server that performs essential SSL interpreting and redirecting the request again to the alternative server.

3. The alternative server gets the demand and checks to see if the consent token given is effective for the requested Fi-ware service. Once verification is successful, the alternative server will send a request to the processing server with the received entity.

4. Before providing the processed object and its matching road segment to the proxy host, the treating host will complete processing and storing the authentic object in Time-scale-DB.

5. Proxy server then obtains the treated entity and forwards the unique demand to the "Orion" agent.

6. The "Orion" Context Agent is responsible for receiving the new entity. An entity, in the context of Orion, represents an object or thing, and new entities might be added to the system based on incoming data. Upon receiving the new entity, the "Orion" Context Agent makes the necessary updates to the MongoDB database. It not only updates the database but also notifies any user who has subscribed to the entity. Subscriptions in this context refer to users expressing interest in receiving updates or notifications related to specific entities or changes in the system.

In summary, the Orion Context Broker serves as a central component in the system, managing the reception and processing of new entities. It ensures that the MongoDB database is updated with the new information, and it also notifies users who have subscribed to relevant entities, providing a real-time and dynamic aspect to the platform. This mechanism supports the idea of providing timely information updates to users who are interested in specific entities within the system.

Users can easily interact and modify facts from the policy. They only need to obtain an effective sign from the "Auth" server, which in turn is used to make appeals to the protected services, such as "Orion," the historical APIs available on the "Auth" server, protected by the alternative server. The "Nginx" server exists between the user and the Auth and Proxy servers, handling SSL decoding and redirecting to the appropriate server. The data flow is mentioned in the below steps:-

1. User-initiated process begins with the user making a request to the Authentication server. The specific purpose of the user's request is to obtain a token. Tokens are used as a form of credentials or proof of authentication that can be presented when accessing protected resources.
2. To obtain the token, the user sends their credentials to the Authentication server. Upon receiving the user's credentials and successfully verifying them, the Authentication server generates a token.
3. A valid token can be used to create a demand for "Orion," which will be processed by the alternative server.
4. The alternative server will check to see if the sign has authorization to create appeals to the corresponding Fi-ware service, and if so, will redirect the request to Orion.
5. The "Orion" Context agent accepts the authentic request, does any necessary operations on Mongo-DB databank, and responds to the alternative server.
6. The alternative server provides the operator with a response that it has received from Orion, i.e., the response is not generated by the Proxy server itself but is a forwarded response from the Orion Context Broker.

When a user wants to obtain historical data, the Auth server's dynamic APIs expose it; consequently, the requirements are not interrupted by the alternative server but are sent openly to the "Auth" host, which in turn validates and executes the requests.

8.5   PLATFORM SCALABILITY

Scalability is a critical consideration in the design and development of various software systems, particularly in the context of modern cloud-based and distributed architectures. It is essential for handling growth, adapting to changing user demands, and ensuring that the system remains responsive and reliable as usage increases. In simpler terms, a scalable platform should be capable of growing and expanding to accommodate higher levels of usage, data, or transactions without causing a significant degradation in performance or a disproportionate increase in resources. Although the TimescaleDB arrangement is effective for the volume of data under consideration, which is not designed to manage a number of devices transmitting data quickly; as more data entered the platform, several dashboard representations became progressively slower. In order to spread the ingestion and query load across numerous servers rather than just one, As a result, we focused on distributed solutions that provide horizontal scaling of database instances, boosting scalability in expressions of communication, storage and processing.

8.5.1 Data Perseverance Scalability

Time-scale-DB supports a mass setup consisting of multiple nodes. In a distributed setup, each node plays a specific role in managing and processing data. The entrance node is a primary instance responsible for controlling and redirecting traffic within the cluster. It likely handles queries, manages connections, and coordinates data retrieval and storage operations. Data knobs are occurrences which stores and operate on the actual data. In a multi-node setup, these nodes work collaboratively to distribute the workload and store data efficiently. Hypertables in TimescaleDB are used to organize and manage time-series data efficiently. They provide an abstraction layer over regular database tables, enabling optimizations for time-series data. Space partitioning on specific columns indicates a strategy for organizing data based on certain criteria beyond just time. This can help in optimizing queries and data retrieval for specific use cases. TimescaleDB inherently supports time partitioning, which involves breaking down large tables into smaller, more manageable chunks based on time intervals. This is particularly useful for time-series data where historical information is often queried differently from recent data. The setup allows for distributing the ingestion process among multiple data nodes. This can lead to improved performance and decreased latency as the workload is distributed, assuming the system is designed to scale horizontally. As the number of nodes increases, indirectly the latency decreases. The emphasis should be done on linear scalability as it proposes a design that enables effective scaling as workload and data volume grow. Apart from this, ClickHouse is a column-oriented DBMS where data is stored in columns rather than rows. This structure is well-suited for analytical queries, where aggregations and analysis are performed on a subset of columns. It has a compression capability which allows the database to reduce the size of stored data, which is beneficial for minimizing storage requirements and potentially improving query performance, especially for analytical queries. Time-scale-DB also offers multi-node architecture, consisting of an entrance node and N number of data nodes, which may be leveraged to maintain the policy as productive as feasible as the number of tenants, data and users grows.

8.5.1.1 Test Facts Generation

The facts are circulated by the fundamental component Fi-ware Orion Framework Agent, which is required by the NGSI paradigm. The proposed model incorporates persistence by storing data in tables with compulsory characteristics: the recognizing object and the time stamp of thoughts and exposes this data through a dynamic API. The API can return data or raw facts with a certain method of sequential aggregation. This data is intended to be shown on a dashboard with the help of composite requests that frequently involve accumulating information.

8.5.1.2   Data Absorption

The CSV data was immediately transferred across every database client, allowing the ingestion rate to be calculated using the biggest batch size available for the assessment data.

8.5.1.3   Query Performance

TimescaleDB is an application that frequently executes a query or series of queries. ClickHouse is described as the fastest for most queries, consistently outperforming TimescaleDB. This performance advantage is noted to be significant in most cases. TimescaleDB is known for its fast execution speeds, especially for queries with lesser amount of rows to aggregate. In such scenarios, TimescaleDB performs well. As the amount of rows that are aggregated increases, TimescaleDB's performance begins to decline practically linearly. This implies that the performance of TimescaleDB is sensitive to the volume of data being processed, and the degradation is noticeable as the dataset size grows. The comparison suggests that the choice between ClickHouse and TimescaleDB might depend on the nature of the queries and the amount of aggregation involved. ClickHouse excels across a broader range of scenarios, while TimescaleDB's strengths lie in situations with lower row aggregation. ClickHouse is presented as an efficient and scalable solution that excels in both query performance and disk space utilization. TimescaleDB, while having a compression feature, faces challenges with complex queries, and the potential solution of continuous aggregates comes with drawbacks related to disk space usage and management complexity.

8.5.1.4   Perseverance Architecture

Even though a single occurrence of Click-House would significantly enhance the policy entirely and simplify the pattern of all service tables, the final aim is to design a dynamic cluster that could be resized at any time based on the needs. As previously stated, both Time-scale-DB and Click-House have methods for constructing a multi-node configuration and together would be competent to cut the latency of more sophisticated requests linearly, as because the amount of requests increased. Following the outcomes, we concluded to adopt Click-House, therefore the subsequent phase is to develop a new design using Click-House and an accessible method. Timescale-DB request, which was implemented on the identical host as of the additional services, was removed and replaced with the appropriate amount of somatic hosts, in which the Click-House and Zoo-Keeper collections will exist in. This collection of somatic hosts is required to be proxy by a weight-paired server, such as "Nginx" server or "chproxy" server created specifically for ClickHouse. It is recommended to employ at least three number of Zoo-Keeper requests, thus we evaluate with minimum of three distinct somatic hosts. Every host then runs the same copy of existing spikes in distinct virtual technologies, allowing all shard copies to be dispersed across multiple somatic hosts. The entire process is significant because complete redundancies would not be achieved if all shard copies were situated on the similar physical device, even if they were in distinct virtual machines. Because each shard maintains a different piece of the distributed table, which is essential for that particular physical system to crash in order to fail access to a portion of the records. If a somatic host fails, one model of all spikes becomes unavailable, but statistics access is not vanishing because there are additional R-1 models accessible for each shard. This policy obviously necessitates a minimum of two duplications per shard, if at least one is wanted while having several shards; moreover, it is recommended to allocate these among separate physical mechanisms in the direction to attain some delicacy, if not totally, as this is not achievable with only copy. This system is active and adaptable to any amount of shards and duplications, requiring just a separate physical host for all copies and a dissimilar virtual mechanism for each shard's copy. In a group of three shards along with five copies, for example, there should be five somatic hosts along with three simulated mechanisms. To add even more redundancy, each instance of ClickHouse might have its own actual machine, which is a costlier method.

The primary goal for the cluster is to enhance availability, and this is expected to be achieved through the use of replicas. Replicas provide redundancy, ensuring that if one node fails, others can still serve the data, thereby improving the overall availability of the platform. Swapping from Time-scale-DB to Click-House with several replicas is expected to significantly speed up the platform. ClickHouse's performance advantages, coupled with the distributed nature of replicas, contribute to faster data access and query execution. Efficient storage can be crucial for managing growing datasets without a proportional increase in storage infrastructure.

8.5.2 Availability of Real-Time Data Transport

This research work on the novel data perseverance architecture can be stretched to other portions of the policy which can be circulated and ready to be changed to original requirements at any moment. Orion is regarded as the platform's central component responsible for distributing information in real time. This indicates that it plays a central role in ensuring that data is disseminated promptly across the platform. During errors, the data is still continued as needed and may be retrieved via the vibrant API; however, real-time information is stopped. It supports high availability setup which involves connecting to MongoDB, a popular NoSQL database. High availability is crucial for ensuring that the system remains operational even in the face of failures. As Orion instances maintain an efficient internal cache of all contributions that are established, any instance can alert a contribution even if it did not get its creation request. The focus on removing single points of failure and maintaining the ability to notify subscriptions across instances contributes to the platform's reliability and performance. Numerous physical mechanisms are utilized to instantiate a particular replica for every Click-House shard. Additionally, an extra simulated machine serves specific functions, by hosting the Orion Framework Agent, the Mongo-DB model, and an Alternative server for request authentication. With this configuration, the multiple Alternative servers that protect the Orion requests must be substituted by a Load Matching server, such as Nginx, which is denoted as one instance in the main host that can also distributed across the remaining of somatic hosts in a distinct virtual mechanism. This decentralized result can be constantly extended to every component of the policy, culminating in a completely redundant solution in which all the single services are duplicated across the existing somatic hosts, in individual and structured virtual machines. Services such as Real-time Server and the MQTT

Connection should have great availability because any one of them fails for any reason, the entire group of Orion agent's stops getting the portion of data. The entire process of providing such facilities will take stretch and must be carefully planned, as difficulties may develop from replica of various services. In this portion, a cluster of physical servers has been established on which we will run our Click-House and Zoo-Keeper clusters, each with a single shard and numerous replicas. Following the migration of all data from Time-scale-DB to the Click-House group, we will begin the process of dispersing platform services among physical servers, beginning with Orion Framework Agent.

8.6 CONCLUSION

This research suggested an accessible platform that can aggregate data from any basis in a smart city and securely and authentically make this data available through various APIs. The architecture of the platform was intended to remain efficient as the volume of data increases and to grow with the data being distributed and ingested. The architecture, which consists of modules that can be distributed and replicated across virtual and physical machines, enables the distribution of the platform load among the hosts and achieves true redundancy. A comprehensive analysis of the persistence options taken into account during the platform's development and implementation was also provided in this article. Although only a few other FIWARE modules are used, FIWARE plays a significant role in the platform as its essential component serves as our main section as well. As a result, the platform handles information persistence and authentication, which were somewhat partial for use cases. However, in the future, if necessary, the policy can be integrated with numerous additional Fi-ware common enablers and components. Future research endeavours to: (1) incorporate additional data sources, including LiDARs, video cameras, and human detection; (2) tackle data fusion; and (3) incorporate predictions into the platform as a whole.

REFERENCES

Ahlgren, B., Hidell, M., Ngai, E.C.H. 2016. Internet of Things For smart cities: Interoperability and open data. *IEEE Internet Computation.* 20 (6): 52–56.

Alberti, A.M., Santos, M.A.S., Souza, R., Da Silva, H.D.L., Carneiro, J.R., Figueiredo, V.A.C., Rodrigues, J.J.P.C. 2019. Platforms for smart environments and future internet design: A survey, *IEEE Access* 7: 165748–165778.

Al-Fuqaha, A., Guizani, M., Mohammadi, M., Aledhari, M., Ayyash, M. 2015. Internet of Things: A Survey on enabling technologies, protocols, and applications, *IEEE Communication Survey & Tutorial.* 17 (4): 2347–2376.

Arora, S., Kumar, M., Johri, P., Das, S., 2016. Big heterogeneous data and its security: A survey, *International Conference on Computing, Communication and Automation, ICCCA.* pp. 37–40.

Badii, C., Bellini, P., Difino, A., Nesi, P. 2020. Smart city IoT platform respecting GDPR privacy and security aspects. *IEEE Access.* 8: 23601–23623.

Bellini, P., Nesi, P., Paolucci, M., Zaza, I. 2018. Smart city architecture for data ingestion and analytics: Processes and solutions, *IEEE Fourth International Conference on Big Data Computing Service and Applications*, pp. 137–144.

Cheng, B., Longo, S., Cirillo, F., Bauer, M., Kovacs, E. 2015. Building a big data platform for smart cities: Experience and lessons from Santander, *IEEE International Congress on Big Data*, pp. 592–599.

Cheng, B., Solmaz, G., Cirillo, F., Kovacs, E., Terasawa, K., Kitazawa, A. 2018. FogFlow: Easy programming of IoT services over cloud and edges for smart cities. *IEEE Internet of Things Journal*. 6 (2): 696–707.

Harris, A., Stovall, J., Sartipi, M. 2019. MLK Smart corridor: An urban test-bed for smart city applications, *IEEE International Conference on Big Data, Big Data, 2019*, pp. 3506–3511.

Hooda, S., Lamba, V., Kaur, A. 2021. AI and Soft Computing Techniques for Securing Cloud and Edge Computing: A Systematic Review, *International Conference on Information Systems and Computer Networks (ISCON)*, pp. 1–5.

Jung, H.S., Yoon, C.S., Lee, Y.W., Park, J.W., Yun, C.H. 2017. Cloud computing platform based real-time processing for stream reasoning, *Sixth International Conference on Future Generation Communication Technologies, FGCT*, pp. 1–5.

Kim, J., Jang, S., Jee, D., Ko, E., Choi, S.H., Kyong Han, M. 2020. 5G based Smart City convergence service platform for data sharing, International Conference on Information and Communication Technology Convergence, pp. 1522–1524.

Kirimtat, A., Krejcar, O., Kertesz, A., Tasgetiren, M.F. 2020. Future trends and current state of smart city concepts: A survey, *IEEE Access*. 8: 86448–86467.

Kolozali, A.E., Bermudez, M. -Edo, Farajidavar, N., Barnaghi, P., Gao, F., Intizar Ali, M., Mileo, A., Fischer, M., Iggena, T., Kuemper, D., Tonjes, R. 2019. Observing the pulse of a city: A smart city framework for real-time discovery, federation, and aggregation of data streams. *IEEE Internet of Things Journal*. 6 (2): 2651–2668.

Lee, J., An, J., Jeong, S., Song, J. 2019. I-LOD: Industrial linked open data system for semantic integration of industrial real-time data in smart city, *IEEE International Conference on Industrial Internet, ICII, 2019*, pp. 297–298.

Lee, R., Jang, R.Y., Park, M., Jeon, G.Y., Kim, J.K., Lee, S.H. 2020. Making IoT data ready for smart city applications, *IEEE International Conference on Big Data and Smart Computing, Big Computing*, pp. 605–608.

Londhe, A., Rao, P.P. 2017. Platforms for big data analytics: Trend towards hybrid era, *International Conference on Energy, Communication, Data Analytics and Soft Computing, ICECDS*, pp. 3235–3238.

Montori, F., Bedogni, L., Bononi, L. 2018. A collaborative Internet of Things architecture for smart cities and environmental monitoring. *IEEE Internet of Things Journal*. 5 (2): 592–605.

Okai, E., Feng, X., Sant, P., Smart cities survey, 2018. *IEEE 20th International Conference on High Performance Computing and Communications; IEEE 16th International Conference on Smart City; IEEE 4th International Conference on Data Science and Systems*, 2018, pp. 1726–1730.

Prashar, N., Hooda, S., Kumar, R. 2023. Current Status of Challenges in Data Security: A Review. In: Jain, R., Travieso, C.M., Kumar, S. (eds) *Cybersecurity and Evolutionary Data Engineering*. ICCEDE 2022. Lecture Notes in Electrical Engineering, p. 1073, Springer, Singapore. https://doi.org/10.1007/978-981-99-5080-5_1

Shahrour, I., Ali, S., Maaziz, Z., Soulhi, A. 2017. Comprehensive management platform for smart cities, *Sensors Networks Smart and Emerging Technologies, SENSET, 2017*, pp. 1–4.

Vítor, G., Rito, P., Sargento, S. 2021. Smart city data platform for real-time processing and data sharing, *IEEE Symposium on Computers and Communications, ISCC*, pp. 1–7.

Wickramasekara, A., Liyanage, M., Kumarasinghe, U. 2020. A comparative study between the capabilities of MySQL and ClickHouse in low-performance linux environment, *20th International Conference on Advances in ICT for Emerging Regions*, pp. 276–277.

Zdraveski, V., Mishev, K., Trajanov, D., Kocarev, L. 2017. ISO-standardized smart city platform architecture and dashboard. *IEEE Pervasive Computation.* 16 (2): 35–43.

Zhang, H., Babar, M., Tariq, M.U., Jan, M.A., Menon, V.G., Li, X. 2020. SafeCity: Toward safe and secured data management design for IoT-enabled smart city planning. *IEEE Access* 8: 145256–145267.

9 From Concept to Reality
The Iterative Path of Smart City Implementation

Gagandeep Kaur, Mandeep Kaur, Righa Tandon, Anisha Singh and Rashmeen Kaur
Chitkara University Institute of Engineering and Technology,Chitkara University , Punjab, India

9.1 INTRODUCTION

Urban planners, technologists and other technologists have been drawn to the idea of smart cities, which provide to enhance residents' quality of life by integrating technology, sustainability, and innovation seamlessly. However, the process of turning this imaginative idea into a practical reality is one of complexity and iteration, marked by difficulties, discoveries and adjustments. Figure 9.1 highlights the use of Internet of things in Smart city applications.

The study examines the development of smart city standards, the important standards that support them, and actual global studies of cities as they navigate the complicated terrain of city transformation in this investigation of the iterative implementation process of smart cities. Smart city drives aren't uniform; The particular qualities, desires and goals of each town might influence them. The smart city concept's adaptability to various contexts is demonstrated by this diversity. Realizing that smart city development is not a one-size-fits-all endeavour is essential as we begin this journey. Rather, it's a unique strategy in which urban communities study from their encounters, repeat their systems, and continually refine their creativity and perception. The accompanying areas will reveal insight into the scopes of brilliant city execution, from ideation to execution, highlighting the challenges confronted, the solutions that were created and the blessings that came along the way. We might also investigate the driving forces behind the smart city movement, which include rapid urbanization, advancements in technology, environmental issues, and an increasing demand for advanced services and infrastructure. These powers have constrained towns and are looking for progressive responses which can adapt to the confounded and interconnected requesting circumstances they face.

In Figure 9.2 percentages are purely sampling and may vary significantly depending on the specific priorities and circumstances of each smart city project. Additionally, we can take a gander at the crucial capability of coordinated effort and commitment to the shrewd town experience. The active participation of numerous stakeholders, including government agencies, private sector partners, academia,

DOI: 10.1201/9781003467892-9

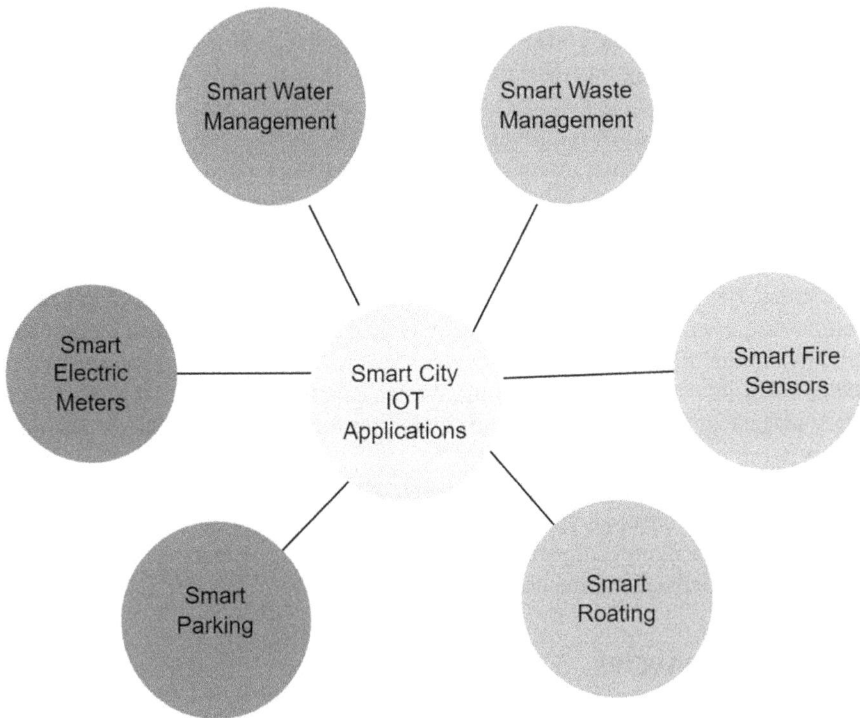

FIGURE 9.1 Use of Internet of Things in smart city applications.

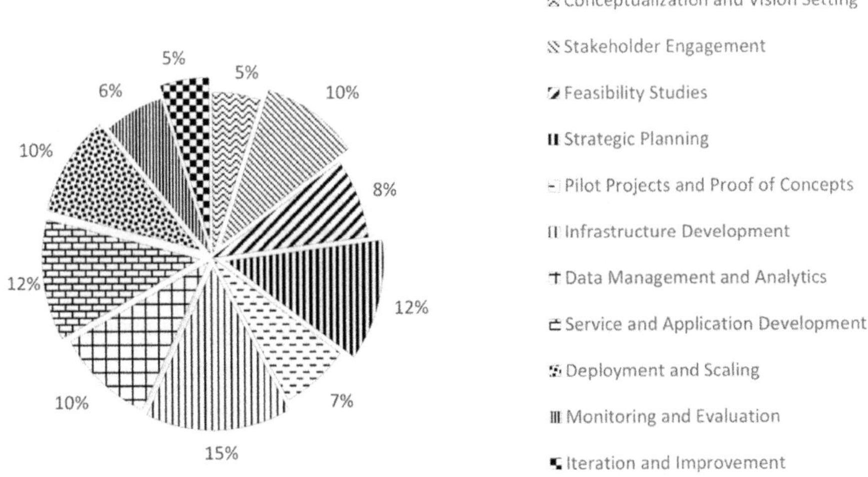

FIGURE 9.2 Areas of the implementation of smart city project.

and, most importantly, citizens themselves, is necessary for successful implementation. Cities can use their collective assets and expertise to transform their urban landscapes by working together. All through this investigation, we will experience case examination and models from urban areas worldwide, exhibiting the different strategies to shrewd town improvement and the preparation they offer. Every case study demonstrates the iterative nature of smart city implementation and the potential for towns to conform and evolve, whether it's the use of statistics analytics to optimize traffic float, the combination of renewable power assets to reduce carbon emissions, or the deployment of IoT sensors to improve public safety.

Ultimately the process of turning an idea into a reality in the world of towns is a demonstration of human ingenuity, creativity, and determination. It is a journey that requires commitment to improving urban lifestyles and visionary management, as well as well-informed decision-making. We can benefit from insights into the changing landscape of city development and its promise for future generations as we move in this iterative direction.

We will focus on the need for government, industry, academia and civil society to work together in developing smart cities concepts. Cooperation and information sharing assume a significant part in defeating difficulties and making the most of the chances that emerge along the way.

9.2 RELATED WORK

Smart-city initiatives driven by the integration of digital technologies to improve urban living have become the focus of contemporary urban development. This literature review examines the re-appropriation of smart cities and explores the journey from concept to reality. The review aims to highlight key issues from scholarly work on strategic improvement, feedback strategies, agile strategies, data-driven decision-making, public and private sector, community participation and adaptive governance in smart city projects and clarifying awareness.

- *Incremental planning options*
  - A review of the literature advocating the benefits of upward planning over massive use.
  - Analysis of case studies demonstrating successful improvement strategies and their impact on project outcomes (Dameri 2017; Fernandez-Aneze et al. 2018).
- *Feedback in urban planning*
  - A review of scholarly work emphasizing the need for continuous feedback in smart city planning.
  - Analysis of methods of data collection from various stakeholders including citizens, businesses and government agencies (Vishnivetskaya & Alexandrova 2019) (Wenge et al. 2014).
- *Accelerated approaches in urban development*
  - An overview of the accelerator methods used in smart city projects.
  - Explore how agile principles help improve flexibility, responsiveness, and productivity (Rao 2018; Dameri 2013; Lopes 2017).

- *Data-Driven Decision-Making in Urban Management*
  - A critical review of the role of data analytics in determining decision-making processes in smart city projects.
  - Evaluating the impact of real-time data on increased urban productivity and efficiency (Syalianda et al. 2021, Drozhzhin et al. 2019).
- *Examples of public-private partnerships*
  - A review of the literature focusing on the benefits and challenges of public–private partnerships in smart cities.
  - Research on successful collaboration models and their contribution to sustainable and innovative smart city solutions (Mohapatra et al. 2021, Falconer & Mitchell 2012, Malik & Shah 2017).
- *Community engagement strategies*
  - An overview of the literature emphasizing the importance of engaging citizens in the planning and decision-making process of smart urban projects.
  - Analysing the use of various community engagement strategies and their impact on project success (Kubina et al. 2021).
- *Adaptive Governance Systems*
  - Assessment of government policies that demonstrate adaptability to ongoing smart city projects.
  - Examining the role of legislative and regulatory flexibility in ensuring effective governance in the integration of emerging technologies.

The importance of an iterative approach was emphasized as a transformative journey from conceptualization to tangible realization of smart city projects. This systematic literature review provides comprehensive insights into the process of reworking in smart city implementation, aiming to bridge the gap between conceptualization and implementation of innovative city solutions between profits (Jayakumar et al. 2021; Hooda et al. 2021).

9.3 THE ITERATIVE APPROACH TO SMART CITY IMPLEMENTATION

The re-implementation approach to smart cities represents a dynamic and adaptable approach that breaks away from traditional top-down, one-time interventions that support the gradual implementation of smart city policies and it's a revolution rather than a massive initiative (Hooda et al. 2021). This approach involves sustainable planning, where stakeholders collectively consider urban needs and challenges. Increased control allows testing and optimization of specific features, reduces the risk of failure and enables the adoption of a customized and efficient approach Process characteristics and responsiveness, which affect about and feedback from the public and stakeholders to assess feasibility and guide future development (Harnal et al. 2022). Data-driven decision-making, agile development processes and open standards ensure that smart urban systems can adapt to respond to changing conditions and emerging technologies Public-private partnerships, community engagement, governance flexibility and resilience and sustainability in the context of complex urban development Contribute further to the success of the iterative approach by fostering This framework recognizes urban challenges and provides continuous learning, adaptation and innovation others prioritize the creation of smart, liveable urban environments (Harnal et al. 2022).

9.3.1 Understanding Iterative Development

The iterative approach is crucial for the successful implementation of smart city projects. Here is what the survey results reveal about understanding conventional development in smart cities:

- Start with a basic IoT-based smart city platform: To ensure scalability and future growth, smart city implementation should start with a basic architecture that is the foundation for integrating other services without compromising business performance (Wenge et al. 2014).
- Consider future applications: getting "smarter" is an ongoing process, and cities should consider not only increasing the number of sensors but also expanding the number of activities at the edge of their smart city also in medicine. For example, a traffic management system can be enhanced to include predictive analytics and traffic signal modulation (Wenge et al. 2014).
- Iterative Process of Design Simulation: In the design simulation phase of smart city development, the initial design is evaluated by simulation to gain insights into various key performance indicators This iterative process helps to refine the design and ensure that it works properly (Rao 2018).
- Advantages for cities of different sizes: Large and small cities can also benefit from the iterative approach. It helps address the high levels of deployment and complexity in urban areas, while in smaller cities it reduces investment in smart solutions and provides constrained infrastructure to use it properly (Wenge et al. 2014; Rao 2018).
- Challenges and planning based on theory: Smart city development requires good planning and implementation of laws, regulations, financial management and ethical standards. For the smart city project to succeed, policy-based challenges such as insight, testing, adoption, heterogeneity and multi-stakeholder engagement must be addressed (Rao 2018).

9.3.2 Why Iteration Matters in Smart Cities

The results of the study indicate that the recycling process is crucial for the successful implementation of smart city projects. Here's why it's important to repeat in smart cities (Kaur et al. 2023a, Kaur et al. 2023b, Sarangi et al. 2022).

- Horizontal efficiency: Smart-city projects are most successful with an iterative, data-driven approach that focuses on pilot projects. While many organizations try to get a return on investment from smart urban initiatives, successful ones say upward mobility works best (Dameri 2013).

- Consider future applications: Getting "smarter" isn't a one-time thing – it's an ongoing process. Cities implementing IoT-based smart city solutions today should consider the infrastructure they would like to implement tomorrow. It envisions an increase not only in the number of sensors but more importantly in the number of operations increased (Dameri 2013).
- Iterative process of design simulation: During the design and simulation phase of smart city development, the initial design is evaluated by simulation to gain insights into various key performance indicators This iterative process helps to refine the design and it is checked for efficiency.
- Advantages for cities of different sizes: Large and small cities can also benefit from the iterative approach. It helps address the high levels of deployment and complexity in urban areas, while in smaller cities it reduces investment in smart solutions and provides constrained infrastructure to use it properly.
- Sustainable approach to design: To reap the benefits of an iterative approach, cities must adopt a sustainable approach to building viable and capable smart cities change.

Overall, the iterative approach helps municipalities stay nimble, gain support, and demonstrate value quickly. It also enables them to deal with the scale and complexity of implementation, reduce investments in smart solutions, and optimize the use of constrained infrastructure resources (Bindal et al. 2019; Reddy et al. 2022).

9.3.3 Iterative Development Lifecycle

The repetition of the developmental life cycle is an important part of smart city implementation. Here's why it's important to repeat in smart cities.

- Incremental approach: Smart city initiatives are increasingly being developed through an iterative, data-driven approach that focuses on pilot projects. This increased access allows for a better return on investment and more efficient use (Lopes 2017).
- Ongoing process: Getting "smarter" isn't a one-time event – it's an ongoing process. Cities implementing IoT-based smart city solutions today should consider the infrastructure they would like to implement tomorrow It aims to increase not only the number of sensors but more importantly the number of operations has gone above.
- Design and Evaluation: During the planning and simulation phase of smart city development, the initial design is tested using simulation to gain insight into key performance indicators of smart city solutions and improve overall performance.
- Functional scalability: Smart city solutions should be designed with functional scalability in mind. This means that existing architecture can easily be updated with new tools and technologies without the need for a complete redesign. The iterative approach allows for this operational scalability, ensuring that smart-city solutions can be adapted for future needs (Lopes 2017).

- Appropriate standards and continuous improvement: In the process of promoting the creation of smart city services, appropriate standards should be set and strictly adhered to according to the overall context of the particular city, development and area is correct. Meanwhile, the ideas, visions, policies and plans, operations and maintenance of smart city infrastructure, development and governance must follow the concept to advance the concept, and technology must follow and review once a year. This continuous improvement is facilitated by an iterative process (Lopes 2017).

9.4   REAL-WORLD CASE

The survey results provide a wealth of real-world data on typical developments in smart cities. Here are some examples:

- Case Studies using Latest technology: For each topic, case studies from cities around the world were combined, including cities that successfully implemented and expanded infrastructure Cities studied the resource had significant leadership and confidence about the local environment. Research has shown that a city that effectively uses data and digital technologies to plan and manage its core services is more efficient, innovative, inclusive and adaptable Digital technologies, especially AI, to drive city services can help it to be more efficient and effective (Jasim et al. 2021; Jayakumar et al. 2021).
- Kansas City, Missouri: The city's innovation partnership program is opening up the city to a three-month pilot project by startups. Other companies submit ideas, and the city chooses the best ones to test in a real-world setting. Bob Bennett, the city's chief innovation officer, said fostering innovation is a cornerstone of the city's IoT strategy, which focuses on big ideas, not technology. Through smart-city projects, the city promotes human-centred innovation (Jasim et al. 2021).

Overall, these real-world scenarios illustrate the importance of conventional improvements in smart cities. The iterative approach achieves a better return on investment, is more efficient, and is more adaptable to future needs.

9.5   THE ROLE OF TECHNOLOGY IN SMART CITIES

Technology has played a significant role in the development of smart cities. Data Correspondence Innovation (ICT) is developed to work on the nature of metropolitan foundation, further develop proficiency and correspondence, diminish expenses and assets, and further develop correspondence among residents and metropolitan occupants. Smart city technologies use their network of connected IoT devices and other technologies to achieve their goals of improving lives and achieving economic growth Successful smart cities follow four steps: resource collection, communication, computing, and control. Smart city planning aims to benefit those people:

residents, businesses and visitors. City leaders should not only raise awareness of the benefits of deployed smart urban technologies but also uphold open and democratic information for their citizens.

By utilizing existing foundation and putting resources into new innovative arrangements, smart urban communities can assist with accomplishing supportable improvement objectives like diminishing emanations, factors that saving, and further developing administrations for residents and organizations As innovation progresses, smart urban areas will turn out to be progressively famous. Shrewd urban communities utilize different advances, like the Web of Things (IoT), huge information, computerized reasoning (artificial intelligence), mechanical technology, distributed computing and 5G organizations.

9.5.1 Technology's Evolution in Smart Cities

Technological advances have played an important role in the development of smart cities. Smart cities use sophisticated technologies and data-driven solutions to improve different aspects of urban life, from infrastructure and transportation to energy and public safety Some potential ways to access technology in cities that here is the wisdom:

1. Information and Communication Technology (ICT): Information technology is a crucial enabler for smart cities. They are utilized to reduce costs and resources, enhance communication between citizens and urban actors, and increase the efficiency, productivity, and collaboration of urban infrastructure.
2. Internet of Things (IoT): The Internet of Things is made up of connected devices that can talk to each other and other systems. Savvy urban communities use IoT gadgets to store information and give ongoing data to work on residents' lives (Syalianda & Kusumastuti 2021).
3. Big data: The vast amount of data generated by IoT devices and other products is analysed using big data. It is used to find improvements, models, and insights that can be used to improve city services and citizens' lives (Syalianda & Kusumastuti 2021).
4. AI: Artificial Intelligence Data analysis and event prediction are two applications of AI. It is used in smart cities to make urban services work better, make traffic flow better, and make.

9.5.2 Data and IoT in Smart Cities

Data and the Internet of Things (IoT) are crucial components of smart cities. IoT sensors are used in cities to collect data and operate systems like traffic management, energy consumption, and waste management. Thus, smart urban communities increment the productivity of metropolitan foundations, decrease costs and convey better expectations of living (Malik & Shah 2017). Here are a few different ways that information and IoT can be created in shrewd urban communities.

1. Data Collection and Analysis: Smart cities collect and analyse data with IoT devices like connected lights, sensors, and meters. This information can be utilized to build moderateness and manageability, work on residents' lives, and make metropolitan administrations more accessible (Kubina et al. 2021).
2. Information and Communication Technology (ICT): Smart cities are driven by software and communication, and they are connected over multiple communication networks. Smart cities use ICT to increase operational efficiency, disseminate knowledge to citizens, and raise standards of public administration and public welfare.
3. Interoperability: City arrangements should be interoperable with the goal that frameworks can cooperate consistently. It uses open protocols and standards to let systems talk to each other and share data (Malik & Shah 2017).
4. Citizen involvement: Citizens and stakeholders should be involved in smart city solutions to make sure that their needs and concerns are met. It involves gathering information and encouraging participation through the use of social media, mobile apps, and other communication channels (Kubina et al. 2021).
5. Real-time feedback: To make people's lives better, smart city solutions need to provide feedback in real time. It draws insights and makes well-informed decisions by utilizing tools for data analysis and visualization (Malik & Shah 2017).

9.6   SUSTAINABILITY IN SMART CITY DEVELOPMENT

A smart city's sustainability is a crucial component. The ultimate goal of smart cities is to enhance the quality of life for their residents by refining the environment in which they live. Coming up next are the absolute most significant maintainability methodologies for savvy city improvement: Sustainable development objectives: aim to reduce carbon emissions, create infrastructure for both individuals and businesses and conserve resources (Saroa & Aron 2018; Kaur & Aron 2022). Technology infrastructure: brilliant urban areas utilize computerized foundation to develop shrewd structures and transport frameworks, and so on. AI and IoT: disruptive technologies that have prevailed in the fight for smart cities: brilliant urban communities depend on shrewd networks to utilize energy effectively and decrease outflows Advanced water frameworks: Smart cities make use of cutting-edge technology to conserve resources while ensuring that people living in cities have access to potable water. Systems driven by data: By improving traffic management and making public transportation more efficient, smart cities work to reduce accidents. Community association: In order to guarantee that citizens and stakeholders are heard and that their requirements are met, smart cities involve them. Utilizing social media, mobile apps, and other channels of communication are all part of this.

9.6.1   Sustainable Urban Planning

Smart cities' efforts to achieve sustainable development objectives depend on sustainable urban planning. These objectives include diminishing fossil fuel byproducts, saving assets, and upgrading administrations for residents and organizations.

A few ways to deal with practical metropolitan arranging can be reliably taken on in the improvement of shrewd urban communities. These include strategies for spatial design that encourage transit-oriented, walkable, and compact cities with mixed-use development and green spaces. Sustainable building methods prioritize environmental friendliness as well as energy and water efficiency. Shrewd vehicle frameworks expect to decrease clog, further develop versatility, and lower emanations through savvy traffic the executives, electric vehicles, and bicycle sharing. Resident interest guarantees that the requirements and worries of residents are thought of, using web-based entertainment and versatile applications for data assembling and empowering contribution. Rain gardens and green roofs are examples of green infrastructure used to manage stormwater, lessen heat island effects, and improve air quality (Verma et al. 2022). Smart cities can improve their sustainability and achieve their development objectives by implementing these strategies.

9.6.2 Environmental Considerations Challenges and Roadblocks

Smart city improvement faces various natural issues, difficulties and barricades. Smart cities should address environmental issues like electricity consumption, carbon emissions, and pollution in addition to pursuing sustainable development goals (Dameri 2017). In smart town improvement, the following are some ability strategies for environmental issues, difficulties and roadblocks (Kaur et al. 2023c).

To achieve sustainable development goals, smart cities need to give priority to energy consumption, carbon emissions, pollution, citizen engagement and interoperability. Smart grids, renewable sources of energy, and energy-efficient structures and organizations should be utilized to address power consumption issues. Sustainable transportation, energy conservation, and the utilization of renewable energy sources all have the potential to lessen carbon emissions. Diminished energy use, green foundation and supportable waste administration are ways of tending to contamination issues. The planning and development of smart cities rely heavily on citizen participation and feedback via social media, mobile apps and other communication channels. Open standards and protocols make interoperability possible, which is required for various systems to seamlessly collaborate. Smart cities can overcome environmental obstacles and achieve sustainable development by implementing these strategies (Tandon & Gupta 2019; Kaur et al. 2023a, b, c; Kumar et al. 2023, Hooda et al. 2021).

9.7 CONCLUSION AND FUTURE SCOPE

The future of smart towns holds the responsibility of more grounded, tolerable, and reasonable metropolitan circumstances. By embracing modern innovation, reality-driven administration, and resident commitment, smart urban areas are able to deal with a variety of complex issues, such as urbanization and climate change. As urban communities proceed to develop and change, the idea of shrewd towns will keep on being significant in making a superior future for the coming ages.

The extent of improvement in the future is promising, with a few important areas of focus. First, improving urban services and citizens' quality of life will require integrating emerging technologies like AI, blockchain, and 5G networks. Second, smart city planning and development will take into account the requirements and concerns of citizens if a citizen-centric approach is utilized. Thirdly, sustainable urban communities that can accommodate growth and change while remaining ecologically sound will require practical metropolitan planning. Fourth, for consistent framework joining, interoperability will be required. Through viable information the executives, brilliant urban areas can at last acquire experiences and pursue all-around informed choices with respect to how to upgrade city administrations and residents' general personal satisfaction. Smart cities can continue to improve their sustainability and achieve their development objectives by implementing these novel strategies.

REFERENCES

Bindal, R., Sarangi, P.K., Kaur, G. and Dhiman, G. (2019). "An approach for automatic recognition system for Indian vehicles numbers using k-nearest neighbours and decision tree classifier." *International Journal of Advanced Science and Technology* 28, no. 9: 477–492.

Dameri, R.P. (2013). "Searching for smart city definition: A comprehensive proposal." *International Journal of Computers & Technology* 11, no. 5: 2544–2551.

Dameri, R.P. (2017). "Smart city implementation." Progress in IS; Springer: Genoa, Italy.

Drozhzhin, S.I., Shiyan, A.V. and Mityagin, S.A. (2019). "Smart city implementation and aspects: The case of St. Petersburg." In *Electronic Governance and Open Society: Challenges in Eurasia: 5th International Conference, EGOSE*, St. Petersburg, Russia, November 14–16, Revised Selected Papers 5, pp. 14–25. Springer International Publishing.

Falconer, G. and Mitchell, S. (2012). "Smart city framework." *Cisco Internet Business Solutions Group (IBSG)* 12, no. 9: 2–10.

Fernandez-Anez V., Fernández-Güell J.M. and Giffinger R. (2018). "Smart City implementation and discourses: An integrated conceptual model. The case of Vienna." *Cities* 78: 4–16.

Harnal, S., Sharma, G., Malik, S., Kaur, G., Khurana, S., Kaur, P., Simaiya, S. and Bagga, D. (2022). "Bibliometric mapping of trends, applications and challenges of artificial intelligence in smart cities." *EAI Endorsed Transactions on Scalable Information Systems* 9, no. 4: e8–e8.

Hooda, S., Lamba, V. and Kaur, A. (2021). "AI and soft computing techniques for securing cloud and edge computing: A systematic review." In *5th International Conference on Information Systems and Computer Networks (ISCON)*, pp. 1–5. IEEE.

Jasim, N.A., Th, H. and Rikabi S.A. (2021). "Design and implementation of smart city applications based on the internet of things." *International Journal of Interactive Mobile Technologies* 15: 4–16.

Jayakumar, J., Nagaraj, B., Chacko, S. and Ajay, P. (2021). "Conceptual implementation of artificial intelligent based E-mobility controller in smart city environment." *Wireless Communications and Mobile Computing* 2021: 1–8. https://doi.org/10.1155/2021/5325116

Kaur, G., Choudhary, P., Sahore, L., Gupta, S. and Kaur, V. (2023a). "Healthcare: In the era of blockchain." In Bipin Kumar Rai, Gautam Kumar, Vipin Balyan (eds.) *AI and Blockchain in Healthcare*, pp. 45–55. Singapore: Springer Nature Singapore.

Kaur, G., Kaur, I., Harnal, S. and Malik, S. (2023c). "Factors and techniques for software quality assurance in agile software development." *Agile Software Development: Trends, Challenges and Applications* 2023: 257–272.

Kaur, M. and Aron, R. (2022). "A novel load balancing technique for smart application in a fog computing environment." *International Journal of Grid and High Performance Computing (IJGHPC)* 14, no. 1: 1–19.

Kaur, V., Kumar, P., Kaur, G. and Kaur, A. (2023b). "Improved facial biometric authentication using MobileNetV2". In *2023 International Conference on Computational Intelligence and Sustainable Engineering Solutions (CISES)*, pp. 376–380. IEEE.

Kubina, M., Šulyová, D. and Vodák, J. (2021). "Comparison of smart city standards, implementation and cluster models of cities in North America and Europe." *Sustainability* 13, no. 6: 3120.

Kumar, A., Hooda, S., Gill, R., Ahlawat, D., Srivastva, D. and Kumar, R. (2023). "Stock price prediction using machine learning." In *2023 International Conference on Computational Intelligence and Sustainable Engineering Solutions (CISES)*, pp. 926–932. IEEE.

Lopes N.V. (2017). "Smart governance: A key factor for smart cities implementation." *IEEE International Conference on Smart Grid and Smart Cities (ICSGSC)*, pp. 277–282.

Malik, F. and Shah, M.A. (2017). "Smart city: A roadmap towards implementation." In *23rd International Conference on Automation and Computing (ICAC)*, pp. 1–6. IEEE.

Mohapatra, H. (2021). "Socio-technical challenges in the implementation of smart city." In *International Conference on Innovation and Intelligence for Informatics, Computing, and Technologies (3ICT)*, pp. 57–62.

Rao, S.K. and Prasad, R. (2018). "Impact of 5G technologies on smart city implementation." *Wireless Personal Communications* 100: 161–176.

Reddy, K., Reddy, M.A., Kaur, V. and Kaur, G. (2022). "Career guidance system using ensemble learning." Proceedings of the Advancement in Electronics & Communication Engineering.

Sarangi, P.K., Sahoo, A.K., Kaur, G., Nayak, S.R. and Bhoi, A.K. (2022). "Gurmukhi numerals recognition using ann." In Pradeep Kumar Mallick, Akash Kumar Bhoi, Paolo Barsocchi, Victor Hugo C. de Albuquerque (eds.) *Cognitive Informatics and Soft Computing: Proceeding of CISC*, pp. 377–386. Singapore: Springer Nature Singapore.

Saroa, M.K. and Aron, R. (2018, December). "Fog computing and its role in development of smart applications." In *IEEE Intl Conf on Parallel & Distributed Processing with Applications, Ubiquitous Computing & Communications, Big Data & Cloud Computing, Social Computing & Networking, Sustainable Computing & Communications (ISPA/IUCC/BDCloud/SocialCom/SustainCom)*, pp. 1120–1127. IEEE.

Syalianda, S.I. and Kusumastuti, R.D. (2021). "Implementation of smart city concept: A case of Jakarta smart city, Indonesia." *IOP Conference Series: Earth and Environmental Science*, 716, no. 1, 012128.

Tandon, R. and Gupta, P.K. (2019). "Optimizing smart parking system by using fog computing." In *Advances in Computing and Data Sciences: Third International Conference, ICACDS*, Ghaziabad, India, April 12–13, Revised Selected Papers, Part II 3, pp. 724–737. Springer Singapore.

Verma, A., Tandon, R. and Gupta, P.K. (2022). "TrafC-AnTabu: AnTabu routing algorithm for congestion control and traffic lights management using fuzzy model." *Internet Technology Letters* 5, no. 2: e309.

Vishnivetskaya A. and Alexandrova E. (2019). "Smart city concept. Implementation practices". *IOP Conference Series: Materials Science and Engineering*, 497, no. 1: 012019.

Wenge R., Zhang X., Dave C., Chao L. and Hao S. (2014). "Smart city architecture: A technology guide for implementation and design challenges." *China Communications* 11, no. 3: 56–69.

10 Trends and Challenges in Development of Smart City and Urbanization System

Mahmood Al-Bahri
Sohar University, Sultanate of Oman, Oman

Wasin Al Kishri
Arab Open University Sultanate of Oman, Oman

10.1 INTRODUCTION

The concept of smart cities emerged due to the convergence of two key movements of our era: the process of urbanization and the advancement of digital technology. The idea of a "smart city" has only been around for a few decades. Despite being initially published in scientific literature in 1994, widespread adoption did not occur until two decades later. The turning point came when the United Nations Economic Commission for Europe Committee on Urban Development and Land Management launched the "United Smart Cities" initiative. This initiative aims to help small towns build an appropriate strategy to use new technologies to achieve economic and social success. With increasing conviction in the feasibility and importance of the opportunities provided by innovation in smart cities, and its positive impact on various aspects of individual and collective life alike and pushing the city towards global leadership by becoming a global investment centre and a pole of discovery and innovation par excellence, with the flow of capital and prosperity that this provides. Economy and increased demand for labour. This chapter discusses the challenges and problems facing smart cities and the Urbanization System, such as security, privacy and electronic threats, pointing out the necessity of confronting these challenges before starting to implement the smart city concept. The chapter will divide the challenges into two parts: the first section discusses the challenges facing smart cities during the embodiment phase, and the second section details the challenges facing smart cities after the embodiment phase.

10.1.1 DEFINITION AND CONCEPT OF SMART CITIES

The notion of a smart city is a relatively recent concept designed to enhance the social and economic dynamics of urban spaces. The purpose of introducing technology is to

DOI: 10.1201/9781003467892-10

improve the living conditions of the urbanized population and solve environmental and demographic problems. Various innovative solutions, most often related to the IT sector, help implement such programs: digital tags, video surveillance systems, centralized public transport, etc.

The saying "The future lies with smart cities" has become widespread in the last ten years. It is often heard from the lips of people involved in urban construction and improvement. Currently, the vast majority of the world's population lives in cities. According to information provided by the UN, by 2050, the urbanization of the population will approach 68%. People prefer to move from small regions to modern megacities (Angel et al. 2011).

However, the increase in population in urban areas leads to significant problems, in particular, traffic congestion, overload of management organizations and other services, increased consumption of energy and other resources, and deterioration of the environmental situation. Administrations are trying to combat these difficulties with the help of information technology. As a result, many large settlements worldwide are becoming intelligent cities or so-called Smart Citiesю. Figure 10.1 illustrates Smart City components.

The concept of smart city is beginning to be used in the world's megacities. However, different countries and companies understand this system differently. For example, among the developers of solutions for smart cities is the IBM organization, which considers the main qualities of a smart year to be its equipment, intelligence and unity. The European Parliament believes that this system's essence is using information and communication resources to solve public problems, including poverty, inequality and unemployment, and obtain better results in energy management (Macrorie et al. 2023).

Despite all the differences in understanding this concept, there is still some commonality in its perception by world states. Smart cities are characterized by

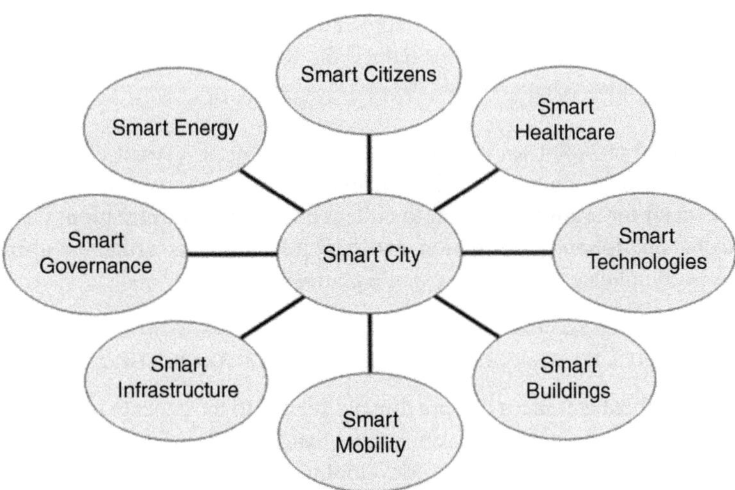

FIGURE 10.1 Smart City components.

sustainability, environmental friendliness, efficiency in data use, and the population's ability to influence management decisions. Their objective is to enhance the quality of life and the level of services for people.

Among the main qualities inherent in smart cities, they should be:

- A strong emphasis on workers, tourists, business owners and the general populace.
- Excellent handling.
- Availability to locals and openness to fresh ideas.
- Unrestricted access to information regarding the actions of the city administration.
- Protection of personal data.
- Integration of services and infrastructure.
- Being proactive in the population's training and development process.

Analytics specialists from Smart Cities Connect claim that in the next 10–15 years, the implementation of the Smart City project will contribute to the progress of most areas of urban life, implying the improvement of the following:

- Public safety. By using crime tracking systems based on video analytics and identifying suspicious sounds and noises, it will be possible to reduce the number of thefts, robbery cases, thefts and murders by 30–40%.
- Social services. By optimizing traffic and street lighting, it can reduce the time it takes for ambulances, firefighters, and police to get to the desired point by 20–35%.
- Resource consumption. It is possible to reduce water and electricity costs by a third and reduce the volume of emissions harmful to the atmosphere by 10–15%.
- Social life. The population will have more opportunities to actively participate in city life with the help of applications.
- Transport system. People can reduce the time to get to work or home by 15–30 minutes (Dameri et al. 2017).

A perfect world presupposes just such an embodiment of a smart city. However, at present, only a few cities have achieved this. According to experts, the main reason is the need for a unified system to collect and analyse information. Data is collected using smartphones, navigation, and search systems. As a rule, the information obtained is complicated to use on a single platform.

10.1.2   Smart City as an Exemplar of Urbanization in the 21st Century

Today's ideas about a modern city are directly related to the concept of a "smart" city, a "digital" city, a "smart city." It seems to us that it will be closest in essence to the organization of life processes of the "developing city" model.

In essence, as gauged by the Human Development Index (HDI), a country's accomplishments in terms of the health, educational attainment, and actual income of its citizens can be categorized into three primary domains:

- Health and longevity, assessed through life expectancy at birth.
- Access to education, evaluated based on the adult literacy rate and the gross enrolment ratio.
- A decent standard of living, measured by Gross Domestic Product (GDP) per capita, plays a significant role in defining the characteristics of a "smart city."

Two polar ones stand out among the implementation of approaches to creating a "smart city" model. At one extreme is the saturation of historically established cities with new functions with the concentration of efforts and resources on developing certain qualities of the urban environment. The other pole involves the creation of "smart cities" in a new location, which is called "from scratch". Both approaches today are of a model nature, focused on practical actions and the development of one or another methodology for the formation of the expected predicted qualities of the urban environment. Fortunately, interest in creating "smart cities" covers ever more expansive areas of activity of the state and society and professionals in various fields of activity.

Several studies by foreign scientific schools, sociological publications and publications focused on innovation are devoted to the issues of sustainable development and the practical application of the smart city model.

The rankings of the most innovative cities in the world include cities that use technological advances to create a favourable, comfortable, safe living environment: Vienna, Toronto, Paris, New York, London, Tokyo, Berlin, Copenhagen, Hong Kong, Barcelona, Boston, San Francisco, Amsterdam, Karamay, Singapore, Songdo. The list of cities shows that these are mainly already established cities with a long development history: New York, Vienna, San Francisco, Amsterdam, Rome, Tokyo. Let us highlight what features experts cite as inherent in a "smart city" using examples of winners' ratings. The fundamentals of a smart city are concentrated in six basic concepts with the prefix "smart": economy, management, residents, movement/mobility, environment, and lifestyle. A few examples: New York is distinguished by well-developed computer systems for city management. More than 60 programs control the operation of public transport housing and communal services in San Francisco, and the world's densest network of public charging stations for electric vehicles has been created here. In Paris, a program of public bicycle parking Velib and Autolib electric vehicles was created with 250 rental stations. Similar programs exist in Munich and other European cities. In Brazil, back in 2006, a resolution was passed on the mandatory equipping of all cars with RFID tags. Information coming from sensors is collected in a single database SYNAV (System for National Identification Automation for Vehicles) (Toh 2022).

Thus, as never before, the global and regional are synthesized into one, giving rise to a qualitatively new environment for human activity in the 21st century.

10.2 TRENDS IN SMART CITY DEVELOPMENT

The smart city development landscape is dynamic, marked by a variety of trends that impact how cities integrate technology and data to enhance urban living. Here is an analysis of some trends in the smart city's development:

10.2.1　IoT and Its Role in Smart Cities

The Internet of Things (IoT) is a relatively new direction actively introduced into Smart Cities. IoT is one of the key technologies that can change modern society and achieve a new level of digitalization.

IoT is a multi-technology concept that denotes that all devices are outfitted with various sensors and linked to the network, enabling remote control, monitoring, and administration of operations in real-time, including autonomous mode. The core idea behind the Internet of Things is an environment where objects can be controlled and where device learning allows data about different objects to be processed to accomplish a task.

It is worth noting that the Internet of Things conceptually belongs to next-generation networks; therefore, its architecture is similar to the well-known four-layer architecture of next-generation networks (NGN). The Internet of Things includes a set of different info-communication technologies that enable the Internet of Things to function, and its architecture shows how these technologies are interconnected. The IoT architecture has four functional layers, as shown in Figure 10.2: IoT architecture.

10.2.2　Cognitive Internet of Things CIoT

IoT is an open paradigm that is highly receptive and adaptive to new principles and architectures related to various directions of development of science and technology. In this regard, using principles and methods of cognition in IoT can be extremely fruitful in creating a cognitive Internet of Things (CIoT).

In addition to being able to perceive information about surrounding objects, CIoT objects can also form specific ideas about the states and operating conditions of those objects, draw logical conclusions from accumulated data, and adapt to both internal and external conditions. In light of this, Figure 10.3: Architecture of Cognitive Internet of Things CIoT or cognitive elements (CE) appear capable of autonomously optimizing, for instance, the technical characteristics of the network under circumstances. On the other hand, CE or CN are merged to form autonomy domains that are part of the Autonomous Domain (AD). These domains are places where devices are relatively close to one another, including within a particular territory, and therefore can cooperate in their behaviour. Each CE or CN will continue to possess the

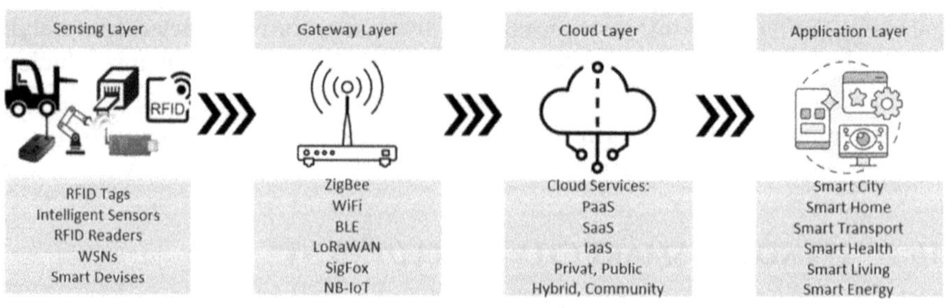

FIGURE 10.2　IoT architecture.

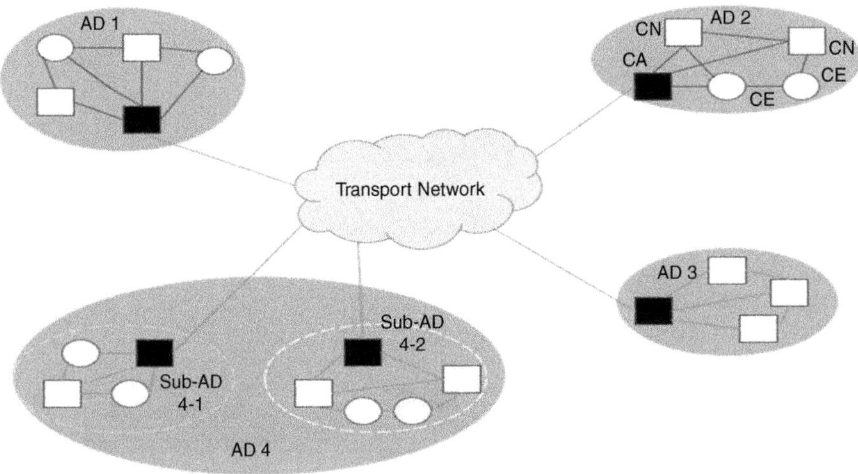

FIGURE 10.3 Architecture of Cognitive Internet of Things CIoT.

property of autonomy in this scenario. Multi-domain cooperation (MDC) allows for the interaction and collaboration of multiple AD domains across international borders. In order to facilitate the organization of such interaction, a cognitive agent (CA) is utilized in every autonomous domain. This CA engages in conversation with the CE or CN that is present in its domain.

Thus, the interaction of domains is possible in general and at the level of an individual cognitive element. Moreover, in each AD domain, there are also superficial, non-cognitive nodes, which are under the control of cognitive nodes (Fan et al. 2023).

10.2.3 Big Data Analytics for Urban Management

The concept of "big data" has become actively used in the commercial sector, scientific literature, and government management documents worldwide over the past ten years. The digitalization of the global economy not only gives rise to new conditions for the everyday life of citizens but also creates new challenges and areas of development for state and municipal governance.

The volume of human knowledge doubles every two to three years, whereas it doubles every century up until the beginning of the 20th century. Today, the volume of human knowledge has doubled at that rate. After the invention of the Internet, 70% of all information that was available was made available. Massive sensor networks are already producing enormous amounts of data streams, which must not only be able to store the data but also process it, draw conclusions from it, and make decisions based on it. This is all while considering that the initial data and processing procedures are frequently inaccurate. "Big Data" is a collection of tools and methods developed in the late 2000s to process large amounts of data. This approach is a collection of tools and methods used to process structured and unstructured data of enormous volumes and significant variety to obtain the

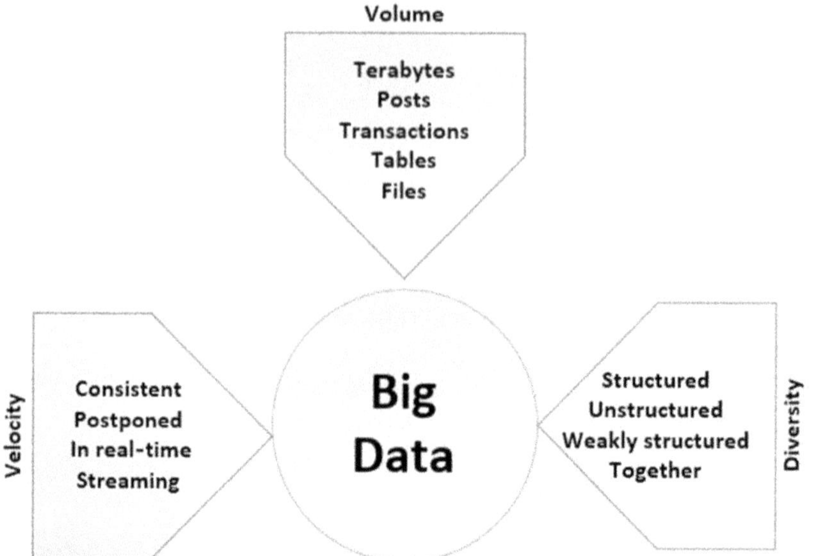

FIGURE 10.4   Three main characteristics of big data.

necessary processing results. Figure 10.4: Three Main Characteristics of Big Data, which explains the three V's of Big D, is summarized as follows: volume (in the sense of the size of the physical volume), velocity (in the sense of both the growth rate and the need for high-speed processing and obtaining results), and diversity (in the sense of the ability to process various types of structured and unstructured data).

The development of Big Data analysis is closely related to the concept of a "smart city." One of the main components of a smart city is the so-called "smart learning," which ensures the collection, storage, monitoring, and analysis of vast amounts of urban data. Big data is the central element of smart learning. On its basis, an analysis of current processes in urban life is carried out, and the accumulated experience is used to adjust further management activities (Khang et al. 2023).

Before describing methods and options for using big data to solve urban problems, it is worth clarifying what this term means. Big data is not just large volumes of information but also a set of operations and procedures for collecting, processing, storing, updating and analysing vast arrays of heterogeneous information. Big Data analysis in public administration focuses on collecting multimodal digital data generated by public and private providers.

Modern technologies make it possible to quickly collect, aggregate and analyse huge amounts of information that can be used in both commercial and non-commercial management processes.

In urban governance, big data can serve as a critical source of statistical information that reflects the needs and concerns of residents in real time. Analysis of domestic and global experience allows us to highlight some of the most critical areas for big data use in city management:

- Analysis of urban transport flows: GPS data on the movement of vehicles allows to develop optimal public transport routes, create new transport hubs, optimize traffic flows, and effectively deal with the problem of traffic congestion.
- Solving communal problems: using citizens' reports about urban issues to solve them quickly.
- Crime control and security: creating a database that combines recordings of city CCTV cameras and citizen reports. In combination with a facial recognition system, the collected data will allow the rapid collection of evidence against persons committing offences. Such information allows quick response to incidents and identifies the city's most disadvantaged areas.
- Response to emergencies or severe weather conditions: Combined operational data can be used to help city authorities take the most effective measures to deal with emergencies and natural disasters, as well as warn the public about danger zones.

In general, it is assumed that the integration of big data of various types, collected both by government agencies and large private organizations, will significantly increase the efficiency of decisions made by city authorities, identify the real needs of residents now, and identify the "weaknesses" of the city's development policy. All these aspects, in turn, can ensure optimization of budget costs.

10.2.4 SMART MOBILITY AND TRANSPORTATION SOLUTIONS

Dynamic transport development poses several global questions for megacities about the further applicability of digital tools for managing transport behaviour, which must be resolved to ensure the sustainability of urban development and the environment to create a comfortable alternative to personal cars, which adversely affect air quality and are the cause of many diseases.

At the moment, a new transport reality is being formed, part of which is gradually becoming trends towards the decarbonization of vehicles, the introduction of car-sharing technologies and alternative modes of transport through the creation of comfortable digital services. The intensive development of technologies such as ML and the IoT, as well as Artificial Intelligence (AI), makes the widespread introduction of unmanned vehicles for personal, public, or shared use possible (Oladimeji et al. 2023).

Presently, urban transport operators worldwide are progressively integrating innovations in infrastructure, rolling stock, and passenger services to address the daily mobility requirements of the population. The array of passenger services has expanded significantly, making it challenging for commuters to navigate and select the most suitable travel option. Addressing this challenge is the contemporary concept of "Mobility-as-a-Service" (MaaS) (Sergio and Wicki 2023), designed to meet the diverse needs of passengers and simplify the process of choosing the most appropriate transportation mode.

Figure 10.5 **Implementation of the MaaS [9]**, shows an example of the implementation of the MaaS concept using one-window technology, which simplifies the passenger's experience in using transport services and passenger services, in

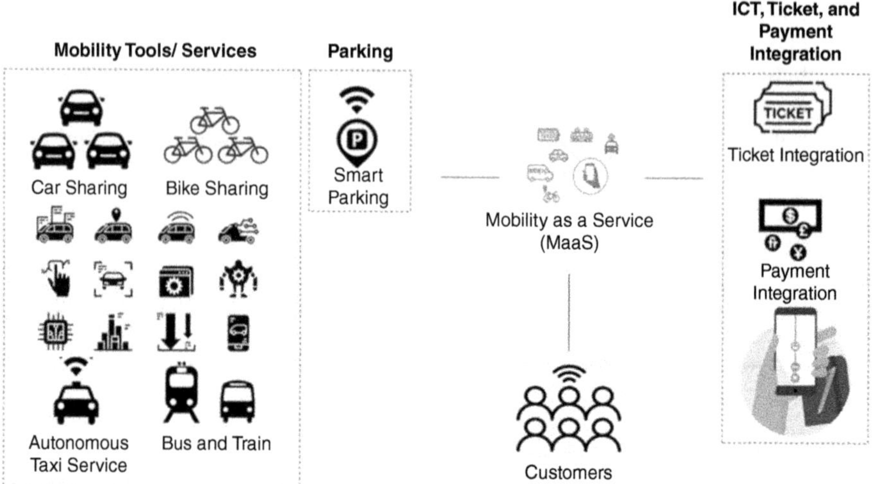

FIGURE 10.5 Implementation of the MaaS.

particular, offers the optimal route and a single method of payment for travel, and also takes into account personal wishes for the trip.

Today, four basic MaaS models are used in megacities around the world:

- MaaS model with the participation of a commercial integrator.
- MaaS model with the creation of an open platform for integration.
- MaaS model with the participation of the city public transport regulator as an integrator.
- Decentralized MaaS model.

Mobility-as-a-Service (MaaS) is a critical component in the evolution of smart cities, playing multifaceted roles to transform urban transportation. By seamlessly integrating diverse modes of transportation and optimizing resources, MaaS enhances the overall user experience, reduces congestion, and lowers emissions. The data-driven insights generated by MaaS platforms aid in informed urban planning, fostering economic efficiency and innovation. Moreover, MaaS contributes to accessibility, inclusivity, and collaborative efforts between public and private sectors, aligning with the holistic vision of creating sustainable and intelligent urban environments.

10.2.5 5G Connectivity

One of the most talked-about subjects in information technology in recent decades is 5G technology. It stands for the ultimate and fifth phase of the development of communications and wireless networks. 5G promises to change industries, revolutionize communication, and pave the way for innovations in a wide range of fields.

5G technology is more advanced than earlier generations due to a number of important features. An important benefit is a notable boost in data transfer speed.

Faster information access makes it possible to develop new high-bandwidth services and applications like virtual reality, cloud computing, streaming high-definition video, and many more.

Moreover, 5G provides a significant reduction in data transmission latency, allowing real-time interactions between networks and devices. This is increasingly crucial in a lot of domains, including industry, entertainment, healthcare and driverless cars. Another major benefit of 5G is more network capacity. This makes it possible for multiple devices to connect to the network at once without compromising performance, which opens the door to the development of smart cities and IoT.

5G technology has a significant impact on several industries in smart cities. It makes it easier for the industrial sector to create smart factories, which improve automation, machine communication in real time, and predictive maintenance. 5G expands access to healthcare services, especially in rural areas, by offering new opportunities for telemedicine, enhanced robotic procedures, and remote patient monitoring. With rich media apps, immersive gaming, and better video streaming, 5G promises to revolutionize the entertainment industry. 5G is essential to the development of autonomous vehicles in the transportation sector because it offers the infrastructure needed for effective data interchange between cars and infrastructure, which enhances overall efficiency, traffic control and safety.

5G technology offers a number of noteworthy advantages and prospects that have a big impact on both life and business. Higher broadband speeds allow for faster data transfers, HD video streaming, and virtual and augmented reality experiences that are immersive. 5G's low data latency makes real-time communication easier, which is especially important for applications like industrial equipment, telemedicine and driverless cars. The Internet of Things is made possible by 5G's expansive network capacity, which opens up new possibilities for industrial automation, smart city development and healthcare. Furthermore, 5G is essential for developing driverless cars, enhancing traffic safety, and increasing the effectiveness of transportation networks. By improving automation, process control, and the application of robotic systems, it also accelerates Industry 4.0 and increases production productivity and flexibility. Additionally, 5G improves entertainment by providing quicker access to virtual worlds, HD streaming, and gaming. These features position 5G as a disruptive force that will be integrated into a wide range of industries, business sectors and our everyday lives.

It is anticipated that 5G networks will keep growing and encroaching on every part of our life in the future. This creates opportunities for technologies that help us build smarter, more integrated cities, industrial complexes, and healthcare systems. 5G-enabled devices are anticipated to become a necessary component of our everyday life.

10.2.6 DIGITAL TWINS

A Digital Twin (DT) is an advanced, integrated simulation of a complex object that incorporates multi-physical, multi-scale, and probabilistic models. It utilizes a combination of physical, mathematical, and simulation models to generate the most precise representation of the actual object. This representation is crafted through the analysis of data collected from sensor networks and various other sources (Qian et al. 2022).

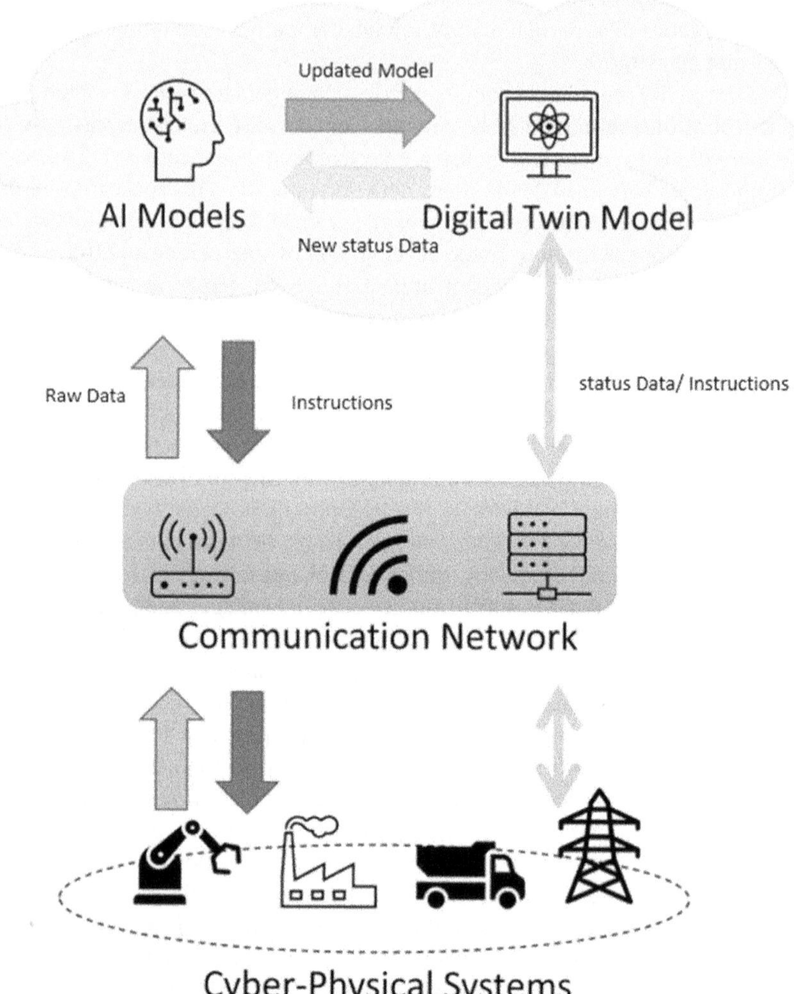

FIGURE 10.6   Architecture of the DT (Khan et al. 2020).

The architecture of a DT for a city is illustrated in Figure 10.6. DT of a city consists of interconnected digital twins that represent different facets of the urban environment. These digital twins enable precise adjustments and synchronization with the real-time condition of urban infrastructure by incorporating data from diverse sources. The seamless functioning of a city's digital twin is contingent on a consistent stream of data generated by various sources within the digital infrastructure of a smart city, as depicted in Figure 10.6.

The collected data ensures the functionality of a network of DT within the city. Examples of such digital twins encompass an interactive 3D model of urban infrastructure, a DT of the transport network for monitoring and forecasting transport accessibility and public transport efficiency, a DT of urban ecology for monitoring and forecasting the ecological state of the urban environment (including soil, water, air quality), and a digital twin of energy, among others.

The creation of a Digital Twin of a city is a complex, comprehensive solution that should be developed incrementally. This involves the integration of private solutions aimed at addressing specific challenges in stages. The adoption of such solutions is not limited to advanced megacities. It is increasingly being implemented in large and medium-sized cities globally. The utilization of these systems, coupled with IoT technologies, enables the resolution of various challenges at a qualitatively different level (Qian et al. 2022).

10.2.7 SMART INFRASTRUCTURE AND BUILDING TECHNOLOGIES

Infrastructures, which are cities, roads, energy, water management, airfields, etc., have been and will be the basis of the economy and people's real lives and work. Their maintenance and development comprise the lion's share of the budgets of both countries and cities. It is utterly unthinkable today to develop them without considering digital transformations and standardized infrastructures and assets, which are mandatory and must meet the most stringent requirements for safety and reliability as the basis of any economy. In the digital economy, the most significant economic advances come from establishing consistent norms and rules across three dimensions: physical, digital and virtual. Recently, the cybernetic world and robots have been added to these three worlds due to the rapid development of this area.

The development of today's standardization also depends on the priorities of the transition to the digital economy. Projects that need to be implemented have severe monetary dimensions and strict deadlines for implementation and have priority in standardization. After the introduction of the concepts of smart city, smart network, smart water, and others into the field of standardization, the era of digital railways and digital production arrived. Expanding the list of such projects is inevitable, and all these areas create their ecosystem of standards, which are reasonably needed to help understand (this is how we are structured). Attempts are made to generalize the phenomena and make them more convenient for perception and understanding.

In infrastructure projects, this evolution gave rise to initial efforts in generalization and the conceptualization of "smart infrastructure." This term denotes the outcome of integrating physical infrastructure with digital infrastructure, delivering enhanced information for decision-making in a more efficient and cost-effective manner.

Infrastructures (both physical and digital) are assets. The introduction of these concepts allows us to consider the main thing – the economic efficiency of specific innovations introduced into practice. It is precisely such financial calculations that make it possible to determine what then refers to explosive technologies in the digital economy, to which we include the great technologies of the 21st century – BIM, GIS, Smart Cities, IoT, robotics and many others (Glaessgen, 2012).

10.3 CHALLENGES IN SMART CITY DEVELOPMENT

Developing smart cities is a complex endeavour marked by intricate challenges spanning technological, social, economic and environmental domains. Challenges such as upgrading water, electricity, and waste systems, especially in older cities with ageing infrastructure, demand innovative solutions. Implementing smart technologies into existing structures poses difficulties, urging a gradual and targeted approach to modernization. The contrast between modern and older cities highlights the need for adaptive strategies, acknowledging the challenges posed by legacy infrastructure.

10.3.1 Infrastructure Challenges: Water, Electricity and Waste Management

Cities face a major problem with their water infrastructure because of outdated systems and pipelines that are prone to leaks and inefficiencies. For effective water distribution, antiquated infrastructure needs to be replaced or repaired. However, there are compatibility issues when integrating smart technology into current networks, necessitating cautious retrofitting and financial outlays.

Upgrading and maintaining water infrastructure may be expensive and disruptive. Smart solution implementation frequently necessitates extensive modification, which can affect both enterprises and households. It is essential to guarantee that people from diverse socioeconomic backgrounds have fair access to clean water. Allocating resources and distributing information strategically are necessary to close the gaps in water supply and quality between neighbourhoods. Getting financing for smart technology integration and infrastructure improvements is a big challenge, particularly for communities with tight budgets. A combination of technological innovation, strategic planning, community engagement and supportive policies is required to construct resilient and sustainable water systems in order to address the issues associated with water infrastructure in smart cities (Kasar et al. 2021).

To ensure effective energy distribution, smart technology must be integrated into electrical grids through upgrades. Technical issues include integrating renewable energy sources, improving grid resilience, and guaranteeing interoperability with smart grid systems. Upgrading and maintaining electrical infrastructure may be expensive and disruptive. Smart solution implementation necessitates extensive refurbishment, which could have an impact on enterprises and inhabitants (Alizadeh et al. 2022).

Implementing efficient waste collection systems, streamlining routes, and ensuring waste is disposed of on schedule are examples of logistical issues. In order to maximize recycling and reduce landfill waste, creative approaches are required. To encourage recycling initiatives and establish a circular economy in metropolitan regions, companies and citizens need to adapt their actions. It is a continuous effort to raise awareness of and participation in recycling activities. The integration of smart technologies for waste monitoring, segregation, and treatment necessitates comprehensive methods. Retrofitting existing waste management systems with sensors and data-driven solutions requires careful planning and money.

In smart cities, problems with waste management, water supply, and electrical infrastructure call for a multifaceted solution. For systems to be robust, effective and sustainable, technological innovation, strategic planning, community involvement, and supportive legislation are essential. To get over these obstacles and create smarter, more sustainable cities, smart grid technology, renewable energy sources, efficient garbage collection techniques, and public awareness campaigns are crucial.

10.3.2 DATA SECURITY AND PRIVACY CONCERNS

The Internet of Things is a network of interconnected systems, gadgets, and sensors found in smart cities. These interdependent parts comprise the backbone of a smart city's functionality, enabling effective operations and data collection.

However, vulnerabilities are also introduced by this interconnection. Data breaches and cyberattacks are more likely in smart city infrastructure because of its heavy reliance on sensors and networked equipment. These networks are interconnected, and hackers might take advantage of them to acquire private information or even take over vital city infrastructure. For example, hackers may disrupt services, create traffic mayhem, or even jeopardize public safety if they manage to access sensors that regulate traffic lights or public transportation systems. Like this, breaking into systems that control the water or electricity supplies could have detrimental effects on how well the city runs. Cybercriminals may also find opportunities in a smart city's massive data collection. If these systems are compromised, data on energy use, traffic patterns, personal information, and other things may be jeopardized (Hamilton et al. 2017).

Because smart cities have so many interconnected systems and collect a lot of data, privacy concerns are important. Massive Data Collection in Smart Cities, for instance, gathers enormous volumes of data from sensors, cameras, and other devices. This contains behavioural data, movement patterns, and personal information. The inhabitants become uneasy knowing that their activities are being watched over constantly and may violate their right to privacy as a result. Furthermore, there may be identifiable information in the data acquired, raising questions regarding personal profiling and possible discrimination based on this information. Furthermore, data security and possible misuse or exploitation are worries when sharing data with outside authorities or service providers.

So, it's imperative to make sure that intrusion detection systems, network segmentation, encryption protocols, and frequent software updates are in place. Strict adherence to privacy laws and procedures also aids in shielding citizen data from misuse or illegal access. To mitigate potential risks and preserve people's privacy, smart city technology should prioritize cybersecurity in addition to ease and efficiency.

10.3.3 FINANCIAL CONSTRAINTS AND FUNDING ISSUES

Large initial outlays are necessary for creating smart city infrastructure, incorporating new technology, and modernizing current services. IoT devices, sensor networks and data analytics platforms are expensive to build and require large sums of capital. It might be difficult to divide up funding for smart city initiatives among competing demands like infrastructure, healthcare, and education.

It can also be difficult to get private sector investment for smart city projects because of unknowns, perceived risks, or a lack of tested business models. Gaining investor trust in smart city projects is challenging since it is hard to produce verifiable proof of their viability and success (Alizadeh et al. 2023).

Innovative financial models, public–private partnerships, sustainable revenue streams and strategic planning are required to meet these financial limits and funding challenges. Stakeholder cooperation and efficient resource management are crucial for overcoming these obstacles and guaranteeing the long-term viability and sustainability of smart city projects.

10.3.4 Social Inclusivity and Digital Divide

In the context of smart cities, social inclusion refers to making sure that all members of the community have equal access to and benefits from the technological advancements and services provided by the smart infrastructure of the city. But achieving social inclusiveness in smart cities presents a big challenge: closing the digital divide.

The disparity between individuals who have access to digital technology – like the Internet and various digital tools – and those who do not is known as the "digital divide." Many factors, including age, education, location, socioeconomic status and others, could contribute to this discrepancy. In smart cities, the digital divide can lead to serious problems and exacerbate socioeconomic inequality (Badran et al. 2023).

Unequal Access to Technology: Some people do not have the same level of access to computers, smartphones, or high-speed internet. The services provided by smart city infrastructure, such as digital healthcare solutions, smart transit systems, and online government services, may not be accessible to some portions of the population. Inequalities in digital literacy can limit people's ability to fully utilize and profit from smart city services, even if they have access to technology. It can be difficult for older people, people from lower-income backgrounds, or people with less education to use digital interfaces and recognize the advantages of technology-driven services.

People without access to digital technologies may be excluded from opportunities in smart cities, including civic involvement, employment, healthcare, and education, due to the digital gap.

Data security and privacy issues: When using smart city systems, people who are not familiar with digital technology may be more vulnerable to privacy violations or the misuse of their personal data. This can further erode their self-assurance and make them steer clear of these services.

10.3.5 Regulatory and Policy Challenges

The creation and execution of smart city projects are severely hampered by regulatory and policy issues. These difficulties frequently result from how difficult it is to incorporate cutting-edge technologies into urban settings while maintaining privacy, security, and moral issues. Among the major issues with regulations and policies are:

- Initially, a multitude of IoT devices in smart cities gather enormous volumes of data. Policies must therefore provide precise guidelines for data collection, storage, sharing and anonymization while guaranteeing the security of personal data.
- Second, there is frequently a lack of system compatibility as a result of several suppliers developing smart city technology independently. Setting up guidelines and conventions for data sharing and communication between various technologies is crucial for smooth integration and effective

operation. Regulations might not be able to keep up with how quickly technologies are developing. It is imperative to establish regulations that are both flexible and adaptable in order to foster innovation and safeguard public safety and interests. In order to address these new issues in cybersecurity, data protection and liability, it will be necessary to amend current laws and regulations.

- Lastly, it is critical to include the public in decision-making processes related to smart city initiatives. In order to build credibility and confidence while introducing new technology, policies should place a high priority on accountability, transparency and citizen involvement (Bhardwaj et al. 2022).

10.3.6 ENVIRONMENTAL SUSTAINABILITY CHALLENGES

Resources like energy, water, and other things are vital to smart cities. It is difficult to strike a balance between the growing need for these resources and environmental objectives. It is essential to manage resources efficiently through the use of technology such as smart grids, water management systems, and waste management solutions. Increased energy consumption may result from the use of smart technologies, such as data centres and Internet of Things devices. Reducing the carbon impact of smart cities requires making sure these systems are energy-efficient and encouraging renewable energy sources (Shang et al. 2023).

It is difficult to design towns that minimize urban sprawl; smart city programs must prioritize green space promotion and efficient transit systems. The establishment of small, mixed-use communities, effective public transportation, and infrastructure that promotes bicycling and walking must be the main goals of these projects. Furthermore, extreme weather and rising sea levels are two effects of climate change that smart cities must be able to withstand. For smart cities to remain sustainable over time, it is imperative to develop systems and infrastructure that can both adapt to and lessen these effects.

10.3.7 CULTURAL AND BEHAVIOURAL CHALLENGES IN ADOPTION OF SMART TECHNOLOGIES

Smart city technology adoption is frequently hampered by behavioural and cultural issues. Cultural resistance to change is one such issue, where people may be reluctant to embrace new technologies because they are accustomed to and attached to customs that are outdated. Another issue that can impede the adoption of smart technologies is the discrepancy in digital literacy and abilities, especially among older generations or marginalized areas. Concerns concerning privacy and trust are particularly important since people could be wary of their personal information being collected and used by smart city technologies. This can create scepticism and resistance towards adopting these technologies. Finally, if individuals do not perceive tangible benefits or utility from smart technologies, they may be less inclined to adopt them. Highlighting the practical advantages, such as convenience, cost savings, or improved services, can encourage adoption (Del Rio et al. 2021, Paroutis et al. 2014).

Analysis of problems in implementing "smart city" programs is necessary to further develop scientific solutions in this direction. First, to create conditions in which some problems will be easily solved, and the emergence of others will become impossible due to the timely adoption of measures to prevent their occurrence. The experience of programs implemented to date, an assessment of the specifics of the countries and cities in which they were launched, will help with the development and modernization of smaller cities, which will contribute to improving the quality of life of their population, subject to competent planning and support for initiatives.

10.4 CONCLUSION

In conclusion, the exploration of trends and challenges in the development of smart city and urbanization systems underscores the dynamic landscape that cities are navigating in the pursuit of intelligent, sustainable, and interconnected urban environments. The integration of big data, IoT, and AI, among other themes, indicates how technology can revolutionize efficiency, safety and urban inhabitants' general quality of life. But problems also arise in tandem with these advances, from cybersecurity risks and privacy issues to the fair application of smart technologies.

It is critical to address these issues as smart cities develop to guarantee resilience, accessibility, and diversity. The dynamic interdependence of urban processes and the swift progress of technology demand an all-encompassing and flexible strategy for urban growth. Overcoming challenges and promoting the sustainable evolution of smart cities will require strong regulatory frameworks, citizen engagement, and joint efforts from the public and private sectors.

The path to smart urbanization is complex and involves constant innovation, thoughtful planning, and a dedication to striking a balance between the growth of technology and the welfare of society. Cities may leverage the promise of smart technology to develop technologically advanced and responsive urban settings that cater to the different demands of their residents by skilfully managing these trends and difficulties. In the end, the emergence of smart cities is evidence of humankind's ability to innovate, adapt, and build urban environments that are more robust and habitable in the face of constant change.

REFERENCES

Alizadeh, Hadi, and Ayyoob Sharifi. 2023. "Toward a societal smart city: Clarifying the social justice dimension of smart cities." *Sustainable Cities and Society*. 95(2023): 104612.

Angel, S., P. Jason, L. C. Daniel, B. Alexander, and P. David. 2011. "The dimensions of global urban expansion: Estimates and projections for all countries, 2000–2050." *Progress in Planning* 75(2): 53–107.

Badran, Ahmed. 2023. "Developing smart cities: Regulatory and policy implications for the State of Qatar." *International Journal of Public Administration* 46(7): 519–532.

Bhardwaj, Neha, Celestine Iwendi, Thaier Hamid, and Anchal Garg. 2022. "Achieving sustainability by rectifying challenges in IoT-Based smart cities." In *International Conference on Big data and Cloud Computing*. Singapore: Springer Nature Singapore. pp. 211–230.

D'Alessandro, Antonella, Hasan Borke Birgin, Gianluca Cerni, and Filippo Ubertini. 2022. "Smart infrastructure monitoring through self-sensing composite sensors and systems: A study on smart concrete sensors with varying carbon-based filler." *Infrastructures* 7(4): 48.

Dameri, Renata Paola. 2017. "Smart city implementation: Creating Economic and Public Value in Innovative Urban System." *Progress in IS*, First Edition. Springer: Genoa, Italy.

Fan, Jiani, Kwok-Yan Lam, and Dusit Niyato. 2023. "Differentiated Security in the Age of Cognitive Internet of Things (CIoT)." *In 2023 IEEE 43rd International Conference on Distributed Computing Systems (ICDCS)*, pp. 965–966.

Glaessgen, Edward, and David Stargel. 2012. "The digital twin paradigm for future NASA and US Air Force vehicles." In *53rd AIAA/ASME/ASCE/AHS/ASC structures, structural dynamics and materials conference 20th AIAA/ASME/AHS adaptive structures conference 14th AIAA*, p. 1818.

Hamilton, Steve, and Ximon Zhu. 2017. "Funding and financing smart cities." *The Journal of Government Financial Management* 66(1): 26–33.

Kasar, Smita, and Meghana Kshirsagar. 2021. "Open challenges in smart cities: Privacy and security." In *Security and Privacy Applications for Smart City Development. First Edition*: 25–36.

Khan, Huma H., Muhammad N. Malik, Raheel Zafar, Feybi A. Goni, Abdoulmohammad G. Chofreh, Jiří J. Klemeš, and Youseef Alotaibi. 2020. "Challenges for sustainable smart city development: A conceptual framework." *Sustainable Development* 28(5): 1507–1518.

Khang, Alex, Shashi Kant Gupta, Sita Rani, and Dimitrios A. Karras. 2023. "*Smart Cities: IoT Technologies, Big Data Solutions, Cloud Platforms, and Cybersecurity Techniques*". CRC Press.

Macrorie, Rachel, Simon Marvin, Adrian Smith, and Aidan While. 2023. "A common management framework for European smart cities? The case of the European innovation partnership for smart cities and communities six nations forum." *Journal of Urban Technology* 30(3): 63–80.

Oladimeji, Damilola, Khushi Gupta, Nuri Alperen Kose, Kubra Gundogan, Linqiang Ge, and Fan Liang. 2023. "Smart transportation: An overview of technologies and applications." *Sensors* 23(8): 3880.

Paroutis, S., B. Mark, and H. Loizos 2014. "A strategic view on smart city technology: The case of IBM Smarter Cities during a recession." *Technological Forecasting and Social Change* 89: 262–272.

Qian, Cheng, Xing Liu, Colin Ripley, Mian Qian, Fan Liang, and Wei Yu. 2022. "Digital twin—Cyber replica of physical things: Architecture, applications and future research directions." *Future Internet* 14(2): 64.

Rio, Del, Dylan D. Furszyfer, Benjamin K. Sovacool, and Steve Griffiths. 2021. "Culture, energy and climate sustainability, and smart home technologies: A mixed methods comparison of four countries." *Energy and Climate Change* 2: 100035.

Sergio, Guidon, Wicki, Michael. 2023. Mobility as a service. (n.d.). Mobility as a Service – Institute of Science, Technology and Policy ETH Zurich. https://istp.ethz.ch/research/mobility/mobility-as-a-service.html

Shang, Dawei, Weiwei Wu, and Daniel Schroeder. 2023. Exploring determinants of the green smart technology product adoption from a sustainability adapted value-belief-norm. perspective. *Journal of Retailing and Consumer Services* 70: 103169.

Toh, Chai Keong. 2022. "Smart city indexes, criteria, indicators and rankings: An in-depth investigation and analysis." *IET Smart Cities* 4(3): 211–228.

11 Trust, Interpretability and Explainability for Ensuring 5G-Enabled Smart City Security

Bikram Kar and Amit Kumar
Sanaka Educational Trusts Group of Institutions,
Durgapur, India

11.1 INTRODUCTION

The development of wireless mobile telecommunications technology occurs in approximately ten-year intervals, defined by the incorporation of fresh frequency ranges, increasing data rates, as well as innovative services that gradually connect our real world. Figure 11.1 depicts the chronological history of mobile communication, highlighting the unique characteristics that accompany each generation. Mobile communication may be traced back to the beginning of the 1980s with the launch of the first generation, also known as 1G. This initial stage laid the groundwork for future advancements in the field. It was notable for its ability to transmit voice using analogue technology. Despite being a significant breakthrough at the time, it had limitations, such being a lack of data services to convert audio transmissions to digital signals, bad voice quality, and a lack of global roaming capabilities. 4G, specifically, offers widespread high-speed wireless internet at the current time; this enables the use of mobile video and cloud applications such as gaming devices, high-resolution mobile streaming and 3D television. At present, 4G technology provides fast data rates for consumers, low latency, and can support about two thousand connected gadgets per square kilometre globally, which helps integrate the Internet of Things. Although 4G has impressive capabilities, the continuous increase in demand and the emergence of new mobile communication technologies suggest that it will likely be surpassed by the upcoming fifth generation (5G) before the next decade begins. The upcoming era of 5G offers unmatched network and service capabilities, bringing forth novel features that have never been seen before. This advancement guarantees uninterrupted network connectivity, improved data speeds, decreased latency, provision for numerous concurrent connections and worldwide availability (Gupta & Jha 2015). Significantly, 5G is ready to tackle the difficulties encountered by the existing 4G, like situations with high mobility (for example, in trains) and areas with varying population densities (such as stadiums and shopping malls). Additionally, 5G will play a crucial role in enabling the true Internet of Things (IoT), acting as a foundation for connecting numerous sensors and actuators while strictly adhering to

DOI: 10.1201/9781003467892-11

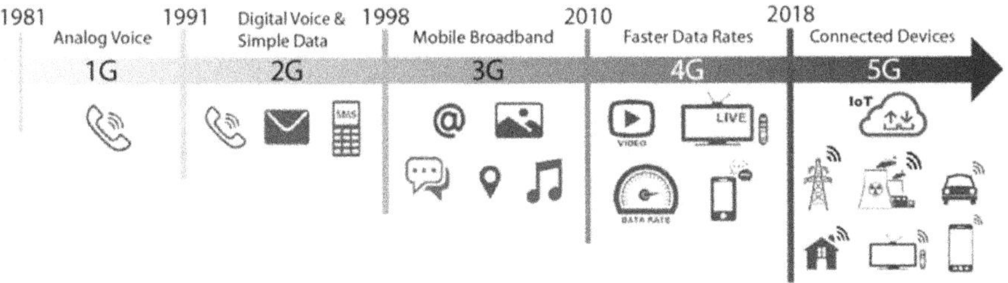

FIGURE 11.1 Mobile Communication Technologies from 1G to 5G Concept of IoT-based smart city (Guevara and Cheein 2020).

transmission limits and energy efficiency (5G-PPP 2020). The progress of mobile communication technologies from 1G to 5G is depicted in Figure 11.1. Driven by the remarkable proliferation of connected devices, an increase in cellphone data traffic, and the limitations of 4G technologies in dealing with this massive data requirement, both the business and academic communities are intensely focused on determining the benchmarks for 5G services, signalling the start of the 5G era. Nowadays, incorporation of 5G technology can improve the performance of Self Driving Vehicles (SDVs) and it will increase the trust of users, manufacturers and stakeholders (Kuru 2022). Blockchain and Deep Learning based technology can be incorporated with 5G to build critical infrastructure (Ragab & Altalbe 2022).

A 5G device can ensure continuous network connection everywhere and at all times, enabling all devices within the network to stay connected. To accomplish this goal, the initial layout of the 5G system is anticipated to handle up to a million connected devices per square kilometre, allowing for the integration of numerous new concepts in IoT services (Technical Report; Samsung Electronics Co). The Internet of Things (IoT) is a modern digital communication concept in which common things can communicate with one another and with persons over the Internet (Zanella et al. 2014). Therefore, the Internet of Things aims to broaden the conventional notion of the Internet, allowing for simple connection with a wide range of devices such as domestic appliances, surveillance cameras, industrial actuators, traffic lights, and automobiles (Technical Report; Samsung Electronics Co 2015). A vast number of interconnected devices produce and gather data on a significant level for a specific goal. The integration of Cloud Computing with Big Data technologies is critical in dealing with various data types, ultimately driving the development of greater value and improved services (Nassar et al. 2018). These types of technologies are crucial for guaranteeing the implementation of the IoT model in urban settings, which are commonly referred to as Smart Cities. They address the requirement for several governments to integrate the use of ICT solutions into public affairs management (Zanella et al. 2014, Yan et al. 2020). 5G technology can be used in smart cities for security and other purposes (Yang et al. 2022). The adoption of 5G technology in intelligent urban areas brings about economic advantages, as well as benefiting specific sectors within smart cities such as energy, transportation, healthcare and entertainment (Ali and Nencioni. 2021). Explainable AI has emerged as a significant contributor in the realm of smart cities, aiming to improve human lives (Javed et al. 2023). XAI, also known as Explainable Artificial Intelligence, has the potential to be utilized in safeguarding against cyber threats, cyber-attacks and unauthorized access to various entities (Kabir et al. 2022). Smart cities aim to enhance the utilization of public resources, improve the quality of services by focusing on convenience, upkeep, and sustainability, and reduce the operational expenses of public utilities within the Internet of Things (IoT) framework (Appio et al. 2019). Latest 5G technology has an important role in upliftment of modern society. 5G technology generates huge data in various forms which promotes Big Data and Intelligent Decision System (Zhang et al. 2023).

11.2 SMART CITIES AND 5G TECHNOLOGY

The notion of Smart Cities has been gaining traction in recent years as cities throughout the world battle with the issues of rapid development, handling of resources, and a need for sustainable growth. This chapter explores the evolution of Smart Cities and delves into the crucial role that 5G technology plays in shaping the landscape of these intelligent urban ecosystems. Figure 11.2 shows an idea of IOT-based smart city.

11.2.1 Evolution of Smart Cities

In response to the rising complexity of urban living, the notion of Smart Cities has evolved. Smart Cities, which were initially motivated by the integration of ICT (Information and Communication Technology), aim to improve inhabitants' quality of life by harnessing technology and information to optimize infrastructure, services and resource utilization. The implementation of sensors, data analysis, and IoT

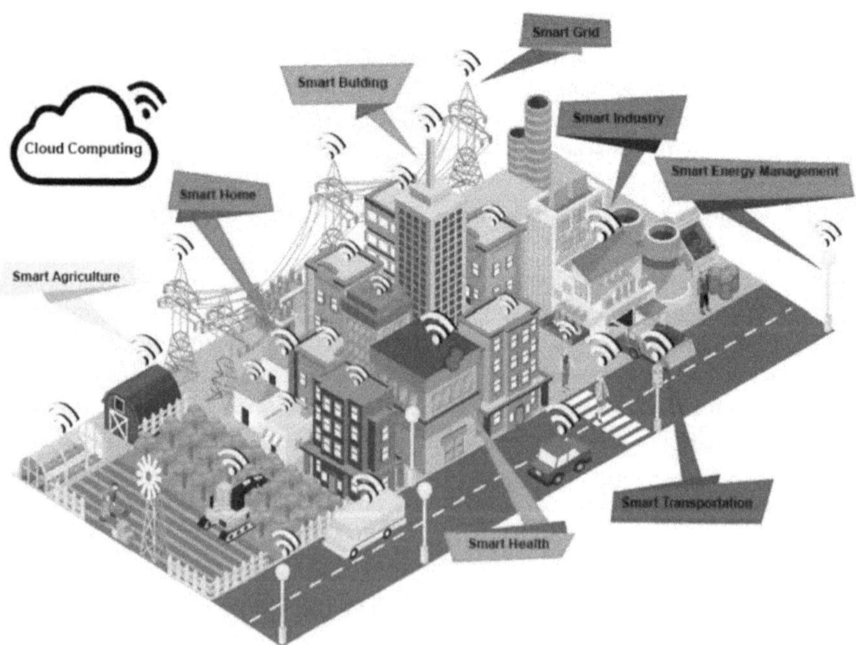

FIGURE 11.2   Concept of IoT-based smart city (Guevara and Cheein 2020).

devices to gather and analyse data from diverse urban systems begins the journey towards Smart Cities. Cities were able to make more informed decisions, increase efficiency, and increase overall sustainability thanks to this data-driven strategy. As technology progressed, the incorporation of 5G represented a watershed moment in the history of Smart Cities.

11.2.2 THE USE OF 5G COMMUNICATIONS TECHNOLOGY IN SMART CITY DEVELOPMENT

5G wireless technology, the fifth generation, provides a quantum advance in communication capabilities. Its high data rates, low latency and widespread connectivity among devices make it a driving force behind the creation of Smart Cities. The use of 5G technology in Smart Cities spans several domains, including:

11.2.2.1 Enhanced Connectivity

5G-enables faster and more reliable internet connectivity, supporting a multitude of devices simultaneously. Improved connectivity is fundamental for the seamless operation of IoT devices, smart sensors and other connected infrastructure.

11.2.2.2 IoT Integration

5G facilitates the integration of a vast number of IoT devices, from smart streetlights and waste management systems to connected vehicles. The low latency of 5G ensures real-time communication between devices, enabling swift responses and enhancing overall system efficiency.

11.2.2.3 Autonomous Systems

By delivering the essential speed and dependability for real-time decision-making, 5G plays a critical role in the advancement of autonomous systems such as self-driving vehicles and drones.

11.2.2.4 Smart Grids and Utilities

The deployment of 5G-enables efficient management of energy grids and utilities through real-time monitoring and control, contributing to sustainability goals.

11.2.3 ADVANTAGES AND DISADVANTAGES OF 5G IN SMART CITIES

While the integration of 5G in Smart Cities brings about transformative benefits, it also presents challenges that warrant consideration.

Advantages:
  i. *High Speed and Low Latency*:
     5G's high data speeds and low latency contribute to faster communication and real-time responsiveness, enhancing the overall efficiency of Smart City systems.
  ii. *Capacity for Massive Device Connectivity*:
     5G can support a vast number of connected devices simultaneously, fostering the growth of IoT applications and interconnected urban infrastructure.

iii. *Innovation and Economic Growth*:
The deployment of 5G in Smart Cities promotes innovation, entrepreneurship and economic growth by creating a fertile ground for the development of new applications and services.
iv. *Improved Public Services*:
Smart City applications powered by 5G, such as smart traffic management and healthcare solutions, contribute to improved public services and a higher quality of life for residents.

Disadvantages:
i. *Infrastructure Costs*:
The implementation of 5G infrastructure requires significant investment, posing a challenge for cities with limited financial resources.
ii. *Security Concerns*:
The increased connectivity in Smart Cities creates potential security vulnerabilities, necessitating robust cybersecurity measures to safeguard against cyber threats.
iii. *Digital Divide*:
The rollout of 5G may exacerbate existing digital divides, as certain communities may struggle to access or afford the technology, leading to disparities in Smart City benefits.
iv. *Health and Environmental Concerns*:
Debates persist about the potential health and environmental impacts of prolonged exposure to 5G radiation, necessitating ongoing research and risk mitigation strategies. The integration of 5G technology in Smart Cities represents a significant leap towards more connected, efficient, and sustainable urban environments. While the advantages are compelling, it is crucial for policymakers, technologists and communities to address the associated challenges to ensure that the benefits of 5G in Smart Cities are inclusive, secure, and sustainable in the long term.

11.3 SECURITY AND PRIVACY CONCERNS IN 5G-ENABLED SMART CITIES

The deployment of 5G technology in Smart Cities heralds a new era of connectivity and innovation, but it also introduces a myriad of security and privacy concerns. As cities become increasingly interconnected through high-speed, low-latency networks, the generation, sharing, and management of vast amounts of data create a complex landscape fraught with potential risks. This chapter explores the security threats, risks and privacy implications associated with the convergence of 5G and Smart Cities.

11.3.1 DATA GENERATION AND SHARING IN SMART CITIES

11.3.1.1 Massive Data Generation

The growth of IoT devices, sensors, and networked systems in Smart Cities leads to large amounts of data being generated. Data has been collected from a variety of sources, including smart infrastructure, public services and citizen interactions, creating a rich but potentially sensitive dataset.

11.3.1.2   Data Sharing Across Systems

Smart City initiatives often involve sharing data across different urban systems to enable seamless functionality and enhance overall efficiency. Interconnected data-sharing ecosystems raise concerns about unauthorized access, data breaches and potential misuse of sensitive information.

11.3.2   SECURITY THREATS AND RISKS

11.3.2.1   Cybersecurity Vulnerabilities

The expanded attack surface in 5G-enabled Smart Cities provides more entry points for cybercriminals. Cybersecurity threats include unauthorized access to critical infrastructure, data breaches, and disruption of essential services.

11.3.2.2   Malware and Ransomware Attacks

The increased connectivity and reliance on digital systems make Smart Cities susceptible to malware and ransomware attacks. An attack on city infrastructure could have severe consequences, affecting public safety and services.

11.3.2.3   Network Slicing Concerns

5G networks implement network slicing to partition the network for different services, introducing potential vulnerabilities if not properly secured. Compromised network slices could lead to service disruptions or unauthorized access to sensitive data.

11.3.2.4   Supply Chain Risks

The global nature of 5G technology supply chains introduces risks related to the integrity of hardware and software components. Supply chain vulnerabilities could be exploited to compromise the security of Smart City networks.

11.3.3   PRIVACY IMPLICATIONS

11.3.3.1   Surveillance and Citizen Tracking

The extensive deployment of sensors and cameras for public safety and traffic management raises concerns about constant surveillance and citizen tracking. Striking a balance between public safety and individual privacy becomes a critical challenge.

11.3.3.2   Data Aggregation and Profiling

The aggregation of diverse datasets in Smart Cities could lead to the creation of detailed profiles of individuals' behaviour and preferences. Unauthorized access to or misuse of aggregated data poses risks to individual privacy.

11.3.3.3   Informed Consent Challenges

Obtaining informed consent for the collection and use of personal data in Smart Cities becomes complex due to the interconnected nature of systems and the sheer volume of data generated. Transparent communication and robust consent mechanisms are essential to address privacy concerns.

11.3.3.4   Lack of Standardization

The absence of standardized privacy practices across Smart City initiatives and 5G implementations may result in inconsistent protection of citizens' privacy. Establishing comprehensive privacy standards is crucial for building trust in 5G-enabled Smart Cities. As Smart Cities leverage the transformative capabilities of 5G technology, it is imperative to address the associated security and privacy concerns to ensure the responsible and ethical development of urban ecosystems. Proactive measures, including robust cybersecurity protocols, transparent data governance frameworks, and privacy-aware design principles, are essential to mitigate risks and build public trust in the evolving landscape of 5G-enabled Smart Cities.

11.4   TRUST AS THE CORNERSTONE OF SMART CITY SECURITY

11.4.1   DIMENSIONS OF TRUST IN SMART CITIES

In the complex and interconnected realm of Smart Cities, trust serves as the linchpin holding together the intricate tapestry of urban innovation, technology and citizen engagement. Trust in Smart Cities can be dissected into several dimensions, each playing a pivotal role in ensuring the security and success of these dynamic urban ecosystems.

11.4.2   NETWORK SECURITY AND TRUST

11.4.2.1   Secure Infrastructure

Trust in Smart City networks hinges on the establishment of secure and resilient infrastructure. The implementation of robust encryption, regular security audits and proactive measures against cyber threats forms the bedrock of a secure network, instilling confidence in its reliability.

11.4.2.2   Standardized Security Practices

Consistent adherence to standardized security practices across Smart City components is imperative. This includes devices, sensors and communication protocols. A uniform security standard provides stakeholders with a reliable framework to assess and trust the security posture of the entire system.

11.4.2.3   Collaborative Security Measures

Building trust in Smart City networks requires a collaborative approach. City authorities, technology vendors, and cybersecurity experts must engage in shared efforts. The exchange of threat intelligence, joint security assessments and a collective commitment to security foster a sense of shared responsibility.

11.4.3   DATA PRIVACY AND TRUST

11.4.3.1   Transparent Data Governance

Trust in Smart City initiatives is contingent on transparent data governance practices. Clear communication regarding data collection and usage policies, obtaining informed consent, and establishing mechanisms for data access and correction contribute to building and maintaining trust.

11.4.3.2 Privacy by Design

Embedding privacy considerations into the design and development of Smart City systems is paramount. Privacy-aware design principles ensure that the protection of personal information is inherent to the technology, cultivating trust among citizens regarding the handling of their data.

11.4.3.3 Regulatory Compliance

Striving to not only meet but surpass regulatory standards for data privacy reinforces trust. Rigorous compliance measures demonstrate a commitment to protecting citizens' privacy and abiding by legal frameworks, thus contributing to the establishment of trust.

11.4.4 USER TRUST AND CITIZEN ENGAGEMENT

11.4.4.1 Inclusive Decision-Making

Fostering trust involves engaging citizens in the decision-making process for Smart City initiatives. Inclusion, transparency, and responsiveness to citizen input contribute to a sense of ownership and trust in the development and evolution of the city.

11.4.4.2 Effective Communication

Open and effective communication channels between city authorities and citizens are essential. Keeping citizens well-informed about Smart City projects, their benefits, and addressing concerns in a timely manner build trust and confidence in the decision-makers.

11.4.4.3 Accessibility and Usability

User trust is reinforced by ensuring that Smart City services are not only technologically advanced but also accessible and user-friendly. Positive user experiences contribute to higher levels of citizen engagement and reinforce trust in the overall Smart City ecosystem.

In the ever-evolving landscape of Smart Cities, where technology and urban living converge, trust emerges as a foundational element for securing the intricate web of connections. By addressing the dimensions of trust in network security, data privacy, and citizen engagement, Smart Cities can cultivate an environment where trust not only safeguards against potential risks but also becomes the catalyst for a symbiotic relationship between the city and its inhabitants, fostering innovation, participation, and a more resilient urban ecosystem.

11.5 INTERPRETABILITY AND EXPLAINABILITY IN SMART CITY SYSTEMS

11.5.1 THE USE OF AI AND ML IN SMART CITIES

As Smart Cities harness the power of AI and ML, these technologies play a vital role in optimizing urban processes, enhancing efficiency and making data-driven decisions. However, the deployment of AI and ML in Smart City systems introduces the critical need for interpretability and explainability to ensure transparency, accountability and user trust.

11.5.2 CHALLENGES IN AI TRANSPARENCY

11.5.2.1 Black-Box Nature of Algorithms

Many advanced AI and ML algorithms operate as black boxes because of its complex nature, making it challenging to understand the decision-making processes within Smart City systems (Hassija et al. 2023). Lack of transparency raises concerns about accountability and the potential for biased or unfair outcomes.

11.5.2.2 Complex Models

The complexity of AI models used in Smart Cities, especially deep learning models, contributes to the difficulty in comprehending their inner workings. Complex models may obscure the factors influencing decision outcomes, hindering interpretability.

11.5.2.3 Data Bias and Fairness

Bias in training data can lead to biased AI outcomes, impacting certain demographic groups unfairly. Understanding and mitigating biases is crucial for ensuring fairness and transparency in Smart City AI applications.

11.5.3 METHODS FOR ACHIEVING INTERPRETABILITY AND EXPLAINABILITY

11.5.3.1 Model-agnostic Techniques

Techniques that are independent of specific machine learning models, such as LIME (Local Interpretable Model-agnostic Explanations), can provide explanations for a model's predictions. These methods offer a post-hoc analysis of model decisions without requiring access to the model's internal architecture.

11.5.3.2 Simplifying Model Architectures

Designing simpler, interpretable models, especially for critical Smart City applications, enhances transparency. While more complex models may provide higher accuracy, simpler models can be easier to understand and explain.

11.5.3.3 Transparent AI Design

Adopting a transparent design philosophy in developing AI algorithms for Smart Cities involves incorporating explainability as a core design principle. Clear documentation of model architecture, training data, and decision-making processes promotes transparency.

11.5.3.4 Explainable AI (XAI) Techniques

Dedicated XAI techniques, such as SHAP (SHapley Additive exPlanations), provide insights into feature importance and contribute to better understanding model decisions. XAI methods aim to bridge the gap between model complexity and interpretability.

11.5.3.5 User-Friendly Interfaces

Implementing user-friendly interfaces that convey AI decisions and predictions in a comprehensible manner fosters trust among end-users. Visualizations, dashboards, and plain-language explanations enhance user understanding of Smart City AI applications.

11.5.3.6 Continuous Monitoring and Auditing

Establishing continuous monitoring and auditing mechanisms for AI systems ensures ongoing transparency. Regular assessments of model performance, bias detection, and explanations for critical decisions contribute to sustained interpretability.

In the evolving landscape of Smart City systems, the integration of AI and ML technologies holds immense promise for innovation and efficiency. However, to garner public trust and ensure responsible AI use, the imperative of interpretability and explainability cannot be overstated. Addressing challenges in AI transparency through diverse methods is not only a technical necessity but also a crucial step towards creating Smart Cities that prioritize accountability, fairness, and user confidence in the age of intelligent urban development.

11.6 PRACTICAL IMPLEMENTATION OF SECURITY MEASURES IN SMART CITIES

As Smart Cities continue to integrate advanced technologies into their infrastructure, ensuring robust security measures becomes paramount to safeguard sensitive data, maintain system integrity, and protect against potential cyber threats. This section explores practical implementations of key security measures in Smart Cities, focusing on encryption for data protection, authentication and access control, and the application of blockchain for data security and integrity. Additionally, real-world case studies highlight successful security implementations in Smart City environments.

11.6.1 ENCRYPTION FOR DATA PROTECTION

11.6.1.1 Implementation

End-to-End Encryption:
   i. Implementing end-to-end encryption ensures that data remains confidential throughout its entire lifecycle, from generation to transmission and storage.
   ii. Strong encryption algorithms, such as AES (Advanced Encryption Standard), are employed to protect sensitive information.

Data-at-Rest Encryption:
   i. Utilizing encryption mechanisms for data stored on servers or devices adds an additional layer of protection.
   ii. File-level or disk-level encryption technologies are applied to safeguard data when it is not in active use.

Benefits:
  i. Mitigates the risk of unauthorized access during data transmission or when stored on devices.
  ii. Protects sensitive information even in the event of a breach, as encrypted data is unreadable without the appropriate decryption keys.

11.6.2 Authentication and Access Control

11.6.2.1 Implementation

Multi-Factor Authentication (MFA):

Implementing MFA requires users to provide multiple forms of identification, such as passwords, biometrics, or one-time codes, to access Smart City systems. MFA adds an extra layer of security beyond traditional username and password combinations.

Role-Based Access Control (RBAC):

Utilizing RBAC assigns specific access privileges based on users' roles within the Smart City ecosystem. This ensures that individuals have access only to the resources necessary for their specific responsibilities.

Benefits:

Strengthens user authentication, reducing the likelihood of unauthorized access. Granular control over access permissions enhances overall system security and minimizes the impact of potential breaches.

11.6.3 Blockchain for Data Security and Integrity

11.6.3.1 Implementation

Decentralized Data Storage:

Utilizing blockchain's decentralized nature for data storage enhances security by eliminating a single point of failure. Data is distributed across a network of nodes, reducing vulnerability to attacks.

Smart Contracts for Secure Transactions:

Implementing smart contracts on a blockchain ensures secure and automated execution of predefined agreements. This is particularly relevant for financial transactions and contractual agreements within Smart City applications.

Benefits:

Enhances data integrity by providing an immutable and tamper-resistant ledger. Increases transparency and trust among stakeholders, as all transactions are recorded in a decentralized and verifiable manner.

11.6.4 Case Studies of Security Implementations

11.6.4.1 Singapore: SecureSmart Platform

Objective: Enhance overall security and resilience in critical infrastructure.

Implementation: Utilizes advanced encryption algorithms, multifactor authentication and blockchain for secure data storage.

Outcomes: Improved data protection, reduced vulnerability to cyber threats and increased resilience in critical infrastructure.

11.6.4.2 Barcelona: BCNNow Platform

Objective: Ensure secure citizen engagement and data privacy.

Implementation: Implements end-to-end encryption for citizen data, employs MFA for authentication, and explores blockchain for secure data sharing.

Outcomes: Enhanced citizen trust, improved data protection, and secure communication channels for city services.

11.6.4.3 Seoul: Blockchain-Based Citizen ID System

Objective: Strengthen citizen identity verification and data protection.

Implementation: Utilizes blockchain for a decentralized citizen ID system, ensuring secure and tamper-resistant identification.

Outcomes: Increased security in citizen identity verification, reduced identity fraud and enhanced data integrity.

The practical implementation of security measures in Smart Cities is essential for fostering trust, protecting sensitive data, and ensuring the overall resilience of urban ecosystems. By incorporating encryption, robust authentication and access control, and leveraging blockchain for enhanced security and data integrity, Smart Cities can navigate the complexities of the digital landscape while safeguarding the interests of citizens and stakeholders. The outlined case studies exemplify successful real-world applications, providing valuable insights for future Smart City security implementations.

11.7 ENSURING TRUST, TRANSPARENCY AND SECURITY IN SMART CITIES

In the dynamic landscape of Smart Cities, ensuring trust, transparency, and security is paramount for the sustainable development of urban ecosystems. This section explores integrated approaches to Smart City security, the importance of stakeholder collaboration and responsibility, and the key policy and regulatory considerations that underpin a secure and trustworthy Smart City environment.

11.7.1 INTEGRATED APPROACHES TO SMART CITY SECURITY

11.7.1.1 Holistic Security Frameworks

Implementation: Develop and implement holistic security frameworks that encompass technological, organizational, and procedural measures.

Benefits: A comprehensive approach guarantees that security considerations are taken into account at all levels, from technology deployment to operational practices.

11.7.1.2 Threat Intelligence and Predictive Analytics

Implementation: Use real-time threat detection and predictive analytics to detect potential security risks before they become serious.

Benefits: Proactive security measures can be implemented to mitigate risks, preventing potential disruptions to Smart City services.

11.7.1.3 Incident Response and Continuous Monitoring

Implementation: To identify and react to safety incidents in real time, establish continuous monitoring tools and comprehensive incident response protocols.

Benefits: Swift responses to security incidents minimize the impact and contribute to the overall resilience of Smart City systems.

11.7.2 STAKEHOLDER COLLABORATION AND RESPONSIBILITY

11.7.2.1 Public–Private Partnerships

Collaboration: Foster partnerships between government entities, private organizations, technology providers, and citizens.

Responsibility: Clearly define roles and responsibilities for each stakeholder in ensuring the security and transparency of Smart City initiatives.

11.7.2.2 Community Engagement

Collaboration: Engage citizens in the decision-making process and involve them in the co-creation of Smart City solutions.

Responsibility: Empower citizens to take an active role in ensuring the security of their data and contribute to the development of secure urban environments.

11.7.2.3 Shared Cybersecurity Resources

Collaboration: Encourage sharing of cybersecurity resources and threat intelligence among Smart City stakeholders.

Responsibility: Collaborative efforts contribute to a collective defence against cyber threats, promoting a culture of shared responsibility.

11.7.3 POLICY AND REGULATORY CONSIDERATIONS

11.7.3.1 Data Protection and Privacy Regulations

Policy: Enact and enforce robust data protection and privacy regulations that govern the collection, storage and use of citizen data.

Regulatory Compliance: Ensure Smart City initiatives adhere to established privacy standards, obtaining necessary consent and protecting individual rights.

11.7.3.2 Standardization of Security Practices

Policy: Advocate for the standardization of security practices across Smart City deployments.

Compliance Requirements: Establish compliance requirements to ensure that security measures are consistently implemented, fostering interoperability and a baseline of security across systems.

11.7.3.3 Regulatory Sandboxes for Innovation

Policy: Create regulatory sandboxes that allow for controlled testing of innovative technologies in a secure environment.

Innovation and Security: Encourage innovation while maintaining a strong focus on security, ensuring that emerging technologies undergo thorough testing before wider deployment.

Ensuring trust, transparency, and security in Smart Cities requires a multifaceted approach that integrates technology, stakeholder collaboration, and regulatory considerations. Holistic security frameworks, collaborative efforts among stakeholders, and robust policy and regulatory frameworks collectively contribute to the creation of Smart Cities that are not only technologically advanced but also secure, transparent and trusted by their citizens. By addressing security challenges through a comprehensive lens, Smart Cities can navigate the complexities of the digital age while prioritizing the safety and well-being of urban communities.

11.8 RECOMMENDATIONS AND FUTURE DIRECTIONS FOR SMART CITY SECURITY

As Smart Cities continue to evolve, it is essential to stay ahead of emerging threats, adopt best practices, and explore new technologies to ensure the ongoing security and success of urban ecosystems. This section provides recommendations for best practices in Smart City security, explores emerging technologies and trends, and suggests areas for future research and development.

11.8.1 BEST PRACTICES FOR SMART CITY SECURITY

11.8.1.1 Conduct Regular Security Audits

Regularly assess and audit Smart City systems for vulnerabilities, ensuring that security measures are up-to-date and aligned with evolving threat landscapes.

11.8.1.2 Implement Zero Trust Architecture

Adopt a zero-trust security approach in which no entity, whether within or outside the network, is trusted by default. This includes continual user and device verification.

11.8.1.3 Prioritize Data Encryption

To secure sensitive information from unwanted access, prioritize the adoption of end-to-end encryption to protect information at rest and in transit.

11.8.1.4 Educate and Train Stakeholders

Conduct training programs to educate Smart City stakeholders, including government officials, citizens and technology providers, on cybersecurity best practices and the importance of maintaining a security-conscious culture.

11.8.1.5 Establish Cross-Sector Collaboration

Foster collaboration between different sectors, including government, academia, private industry, and citizens, to share knowledge, resources and threat intelligence.

11.8.2 EMERGING TECHNOLOGIES AND TRENDS

11.8.2.1 AI-Driven Security Analytics

Leverage artificial intelligence for advanced threat detection and security analytics, enabling Smart Cities to identify and respond to security incidents in real time.

11.8.2.2 Quantum-Safe Encryption

Anticipate the future threat posed by quantum computing to current encryption methods by adopting quantum-safe encryption algorithms.

11.8.2.3 5G Security Enhancements

Continuously explore and implement security enhancements specific to 5G networks, addressing potential vulnerabilities and ensuring the integrity of Smart City communications.

11.8.2.4 Edge Computing Security

Strengthen security measures for edge computing, ensuring that data processed at the edge remains secure, and edge devices are protected against cyber threats.

11.8.2.5 Resilient IoT Security

Develop and implement security measures that enhance the resilience of IoT devices, protecting them from compromise and ensuring the integrity of Smart City IoT ecosystems.

11.8.3 FUTURE RESEARCH AND DEVELOPMENT

11.8.3.1 Privacy-Preserving Technologies

Explore and develop advanced privacy-preserving technologies, such as homomorphic encryption and secure multi-party computation, to safeguard citizen privacy in Smart City data processing.

11.8.3.2 Autonomous Security Systems

Investigate the development of autonomous security systems that use AI to autonomously detect, respond to, and mitigate security threats without human intervention.

11.8.3.3 Blockchain for Identity Management

Research the use of blockchain for decentralized and secure identity management systems, enhancing the security of citizen data and authentication processes.

11.8.3.4 Cyber-Physical System Security

Focus on the security of cyber-physical systems within Smart Cities, including smart grids, transportation systems, and healthcare infrastructure, to ensure the resilience of critical services.

11.8.3.5 Human-Centric Security Design

Explore human-centric design principles for security solutions, ensuring that security measures are user-friendly, easily understandable, and promote a positive user experience.

As Smart Cities embrace innovation and connectivity, the security landscape must evolve in tandem. Implementing best practices, adopting emerging technologies, and investing in ongoing research and development will fortify Smart City security against emerging threats. By staying proactive and collaborative, Smart Cities can build resilient and secure urban environments that prioritize the well-being of citizens and the sustainable growth of urban ecosystems.

11.9 CONCLUSION

In conclusion, the trajectory of Smart Cities, shaped by the integration of technologies like 5G, AI, and IoT, necessitates a meticulous focus on security and a foundation built on trust and transparency. The journey from the early stages of information technology integration to the current comprehensive ecosystems leveraging cutting-edge technologies underscores the potential for enhanced efficiency, sustainability and overall quality of life. However, as the cities of the future become increasingly interconnected and data-centric, concerns surrounding security and privacy come to the forefront. The deployment of 5G technology, while promising unprecedented connectivity and speed, brings forth the imperative to navigate challenges and embrace opportunities with a keen eye on ensuring the security and privacy of citizens. Trust emerges as a pivotal factor in this equation, permeating through institutional, technological, and interpersonal dimensions. The role of interpretability and explainability in AI applications becomes crucial, not just for addressing concerns but for fostering ethical and responsible development practices in Smart City initiatives. In this context, practical security implementations such as encryption, authentication and blockchain serve as tangible safeguards fortifying Smart City systems against potential threats. Holistic security frameworks, continuous monitoring and collaborative efforts among stakeholders provide the necessary resilience for these urban ecosystems. Stakeholder collaboration, including public–private partnerships and community engagement, establishes a collective responsibility for the security of Smart City initiatives. Policy considerations, from enforcing data protection and privacy regulations to standardizing security practices, shape the regulatory landscape crucial for the secure evolution of Smart Cities. As these cities increasingly rely on 5G technology, the path forward involves proactively addressing challenges, maintaining transparent communication, and engaging in collaborative efforts to ensure a future that is not only technologically advanced but also deeply rooted in trust, resilience, and the well-being of citizens.

REFERENCES

5G-PPP. The 5G Infrastructure Public Private Partnership: The Next Generation of Communication Networks and Services. (White Paper). Available online: https://5g-ppp. eu/wp-content/uploads/2015/02/5G-Vision-Brochure-v1.pdf (accessed on 21 July 2020).

Ali, Gohar, and Gianfranco Nencioni. "The role of 5G technologies in a smart city: The case for intelligent transportation system." *Sustainability* 13, no. 9 (2021): 5188.

Appio, Francesco Paolo, Marcos Lima, and Sotirios Paroutis. "Understanding Smart Cities: Innovation ecosystems, technological advancements, and societal challenges." *Technological Forecasting and Social Change* 142 (2019): 1–14.

Guevara, Leonardo, and Fernando Auat Cheein. "The role of 5G technologies: Challenges in smart cities and intelligent transportation systems." *Sustainability* 12, no. 16 (2020): 6469.

Gupta, Akhil, and Rakesh Kumar Jha. "A survey of 5G network: Architecture and emerging technologies." *IEEE Access* 3 (2015): 1206–1232.

Hassija, Vikas, Vinay Chamola, Atmesh Mahapatra, Abhinandan Singal, Divyansh Goel, Kaizhu Huang, Simone Scardapane, Indro Spinelli, Mufti Mahmud, and Amir Hussain. "Interpreting black-box models: A review on explainable artificial intelligence." *Cognitive Computation* (2023): 1–30.

Javed, Abdul Rehman, Waqas Ahmed, Sharnil Pandya, Praveen Kumar Reddy Maddikunta, Mamoun Alazab, and Thippa Reddy Gadekallu. "A survey of explainable artificial intelligence for smart cities." *Electronics* 12, no. 4 (2023): 1020.

Kabir, M. Humayun, Khondokar Fida Hasan, Mohammad Kamrul Hasan, and Keyvan Ansari. "Explainable artificial intelligence for smart city application: a secure and trusted platform." In *Explainable Artificial Intelligence for Cyber Security: Next Generation Artificial Intelligence*, pp. 241–263. Cham: Springer International Publishing, 2022.

Kuru, Kaya. "TrustFSDV: Framework for building and maintaining trust in self-driving vehicles." *IEEE Access* 10 (2022): 82814–82833.

Nassar, Ahmed Samy, Ahmed Hossam Montasser, and Nashwa Abdelbaki. "A survey on smart cities' IoT." In *Proceedings of the International Conference on Advanced Intelligent Systems and Informatics 2017*, pp. 855–864. Springer International Publishing, 2018.

Ragab, Mahmoud, and Ali Altalbe. "A blockchain-based architecture for enabling cybersecurity in the internet-of-critical infrastructures." *CMC-Computers, Materials & Continua* 72 (2022): 1579–1592.

Samsung. 5G Vision (White Paper); Technical Report; Samsung Electronics Co.: Suwon-si, Korea, 2015.

Yan, Jianghui, Jinping Liu, and Fang-Mei Tseng. "An evaluation system based on the self-organizing system framework of smart cities: A case study of smart transportation systems in China." *Technological Forecasting and Social Change* 153 (2020): 119371.

Yang, Chen, Peng Liang, Liming Fu, Guorui Cui, Fei Huang, Feng Teng, and Yawar Abbas Bangash. "Using 5G in smart cities: A systematic mapping study." *Intelligent Systems with Applications* 14 (2022): 200065.

Zanella, Andrea, Nicola Bui, Angelo Castellani, Lorenzo Vangelista, and Michele Zorzi. "Internet of things for smart cities." *IEEE Internet of Things Journal* 1, no. 1 (2014): 22–32.

Zhang, Denghui, Zhaoquan Gu, Lijing Ren, and Muhammad Shafiq. "An interpretability security framework for intelligent decision support systems based on saliency map." *International Journal of Information Security* (2023): 1–12.

12 Enhancing Project Security

Unveiling Trust, Interpretability and Explainability in the Age of AI

Marwan Alshar'e, Abdallah Abualkishik, Khaled Abuhmaidan and Ahmad Kayed
Sohar University, Sohar, Oman

12.1 INTRODUCTION BACKGROUND AND SIGNIFICANCE

The incorporation of artificial intelligence (AI) into many fields has brought about significant potential for change, but it has also introduced intricate difficulties, notably in the sphere of security. The first paragraph establishes the foundation for a thorough investigation of the significance of trust, interpretability and explainability in guaranteeing the security and ethical implementation of projects powered by artificial intelligence.

The widespread use of artificial intelligence (AI) technology has resulted in remarkable progress in several sectors, including healthcare, finance and autonomous systems. Nevertheless, the rapid advancement of technology raises significant security issues that need a comprehensive comprehension of how trust, interpretability and explainability may strengthen the fundamental aspects of artificial intelligence initiatives.

12.1.1 RESEARCH OBJECTIVES

The main aims of this chapter are to explore the complex relationship between trust, interpretability, and explainability, clarifying their individual and combined contributions to enhancing the security of projects powered by artificial intelligence. By integrating many sources of literature, real-world instances and prospective viewpoints, our objective is to provide valuable insights that may guide the development of safe AI systems and contribute to the continuing scholarly conversation on this topic.

DOI: 10.1201/9781003467892-12

12.1.2   Scope of the Study

This research undertakes a comprehensive examination of trust, interpretability and explainability within the domain of AI project security. This study examines the practical uses, obstacles, and prospective developments in several fields, including healthcare, finance and autonomous systems. The breadth of the subject matter expands beyond just technical issues to embrace a broader range of ethical consequences and regulatory systems.

12.1.3   Methodology

The technique used encompasses an extensive survey of relevant scholarly literature, a meticulous analysis of practical case studies, and a prospective evaluation of developing trends. This chapter seeks to provide a comprehensive viewpoint on the incorporation of trust, interpretability, and explainability in order to provide strong security for AI projects. This will be achieved by combining ideas from academic research, industry practices and ethical issues.

12.1.4   Motivation for the Study

The expeditious advancement of artificial intelligence (AI) technology has given rise to a multitude of prospects and complexities. The rationale for doing this research arises from the need to effectively tackle the urgent requirement for a comprehensive comprehension of how trust, interpretability, and explainability might alleviate security threats in artificial intelligence (AI) endeavours. The necessity to explore the complex dynamics of AI components is highlighted by recent developments in the field, as well as instances of security breaches and ethical problems. This research aims to provide a meaningful contribution to the ongoing conversation on ensuring the future security of AI-driven technology by identifying and resolving gaps in current understanding.

12.2   THE ROLE OF TRUST IN AI SYSTEMS

There is a significant concern over information security assaults and the need for safe systems in distributed contexts, where threats might originate from several sources. Trusted computing is a concept that was developed to solve consumers' worries about data security while connected to a network. Its goal is to inspire trust by protecting data, maintaining the reliability of the platform and enabling users to make educated choices about trusting other networks (Alshar'e et al. 2014).

Trust plays a crucial role in the effective integration and implementation of artificial intelligence (AI) systems, exerting a substantial influence on their design, use and public perception. This section delves into the many dimensions of trust in artificial intelligence (AI), highlighting its relevance in the realm of technology, its function as a fundamental element of project security, and the necessity of openness in fostering confidence in AI systems.

12.2.1　The Importance of Trust in Technology

The notion of trust is inherently interconnected with technology, specifically within the domain of artificial intelligence. As artificial intelligence (AI) systems continue to proliferate in many aspects of our everyday routines, both people and enterprises demonstrate a significant level of reliance on these technological advancements (Smith et al., 2020). Trust, within the given context, pertains to the confidence placed in AI systems to function consistently and make judgments that are in accordance with the intended objectives. The trust mentioned here encompasses both consumers who depend on AI-driven personal assistants and organizations that use AI for their decision-making processes.

12.2.1.1　Building Trust with Ethical AI Design

In order to establish trust, it is crucial for developers of artificial intelligence (AI) to give utmost importance to ethical design principles. The use of ethical principles in the usage of artificial intelligence (AI), such as fairness, accountability, transparency, and interpretability, has been emphasized by (Floridi et al., 2018). This emphasis aims to guarantee that AI systems are developed in a manner that upholds essential ethical standards, thereby encouraging trust among both users and stakeholders.

12.2.2　Trust as the Foundation of Project Security

In the context of AI initiatives, trust plays a pivotal role not only in shaping user perception but also in ensuring the security of the project. The domain of project security comprises the measures taken to ensure the safety of data, the mitigation of vulnerabilities and the prevention of hostile assaults. Trust plays a crucial role in the attainment of these security goals.

12.2.2.1　Trust-Centric Security Measures

The adoption of a trust-centric framework for project security entails acknowledging the inherent interdependence between trust and security. Enhancing security may be achieved by placing a priority on trust throughout the whole process of developing, deploying and maintaining the AI system. Trust-centric security measures include a range of strategies aimed at ensuring the integrity and reliability of an AI system. These measures often involve robust data protection protocols, secure model training procedures and ongoing monitoring of the system's behaviour (Doshi-Velez et al., 2017).

12.2.3　Transparency and Trust in AI Systems

The establishment of transparency plays a crucial role in fostering confidence in artificial intelligence (AI) systems. Transparency pertains to the degree to which the internal mechanisms, decision-making procedures and data management practices of an AI system are disclosed and understood by users and those with a vested interest.

12.2.3.1 The Role of Explainable AI (XAI)

The use of Explainable AI (XAI) methodologies is of paramount importance in enhancing transparency. These methodologies facilitate users in comprehending the rationale behind a specific choice made by an artificial intelligence (AI) system, hence mitigating the often observed notion of opaqueness associated with intricate AI algorithms (Rudin, 2019).

12.2.3.2 The Regulatory Landscape

The significance of openness in AI systems is being acknowledged by governments and regulatory agencies. The deployment of artificial intelligence (AI) is subject to transparency standards as outlined in legislative measures such as the General Data Protection Regulation (GDPR) of the European Union (European Union, 2016/679). Additionally, developing rules specifically targeting AI also underlines the need for openness in companies using this technology. The adherence to these standards not only amplifies the level of openness but also fortifies the level of confidence in artificial intelligence systems.

Within this particular section, we have thoroughly examined the pivotal significance of trust within artificial intelligence (AI) systems. We have highlighted its value within the realm of technology, accentuating its position in ensuring project security. Furthermore, we have engaged in a comprehensive discussion about the intricate link that exists between transparency and trust. The use of in-text citations and references serves to substantiate the material and provide readers with additional sources for further investigation of the subjects addressed.

As we explore the complex terrain of artificial intelligence, it becomes clear that trust is the crucial element that connects the domains of technology, security and transparency. The intricacies examined in this part emphasize that trust is not just a subjective feeling of the user, but rather a crucial and essential aspect of strategy. The fundamental significance of trust is seen in the convergence of ethical AI design, transparent behaviours, and security measures. This interaction encourages more investigation into how trust influences moral foundations and security measures in the evolving field of artificial intelligence.

12.3 THE CONCEPT OF INTERPRETABILITY

The idea of interpretability has significant importance within the field of artificial intelligence, since it has far-reaching consequences for the dependability, safeguarding and adoption of AI systems by users. This section explores the notion of interpretability, highlighting its significance in the field of artificial intelligence (AI), its function in ensuring the security of AI systems, and its capacity to act as a link between opacity and understanding.

12.3.1 UNDERSTANDING INTERPRETABILITY IN AI

12.3.1.1 Defining Interpretability

The concept of interpretability in the field of artificial intelligence pertains to the degree to which the behaviour and decision-making mechanisms of an AI system may be comprehended, elucidated, and rationalized by individuals, namely end-users and stakeholders (Lipton, 2016). Explainability refers to the comprehensibility of the internal mechanisms of an artificial intelligence (AI) model, as well as its capacity to provide elucidation on the rationale behind a particular choice.

12.3.1.2 Dimensions of Interpretability

Interpretability encompasses several manifestations, such as the transparency of models, the significance of features, the logic behind decisions and the consistency of explanations (Doshi-Velez & Kim, 2017). The comprehensibility of an AI system is contingent upon the aspects outlined above, which govern the extent to which humans may effectively engage with the system.

12.3.2 ROLE OF INTERPRETABILITY IN SECURITY ASSURANCE

12.3.2.1 Enhancing Trustworthiness

The inclusion of interpretability is of utmost importance in ensuring the security of AI applications. Organizations may strengthen the trustworthiness of AI systems by increasing their interpretability, which in turn helps to mitigate the potential risks associated with malicious assaults and inadvertent mistakes (Rudner & Toner, 2021).

12.3.2.2 Detecting Adversarial Attacks

AI models that are interpretable have superior capabilities in detecting and effectively countering adversarial assaults. According to (Papernot et al., 2017), the identification of abnormalities and unexpected behaviours becomes more feasible when the decision-making process of an AI system is transparent.

12.3.3 BRIDGING THE GAP BETWEEN OPAQUENESS AND COMPREHENSION

12.3.3.1 The Challenge of Opaqueness

Numerous sophisticated artificial intelligence (AI) models, like deep neural networks, possess an inherent opacity that poses challenges in comprehending their fundamental mechanics. According to (Chen et al., 2018), the presence of opaqueness might impede the development of trust and give rise to potential security vulnerabilities.

12.3.3.2 Techniques for Bridging the Gap

Efforts aimed at reducing the disparity between the lack of transparency and the capacity to understand AI systems include the advancement of interpretable AI models, strategies for explainable AI (XAI), and approaches for interpretability that are not dependent on specific models (Ribeiro et al., 2016). The aforementioned methodologies strive to provide valuable perspectives on the behaviour of artificial intelligence systems, especially in cases when the underlying models exhibit complexity.

By analysing the complexities of interpretability in AI, we reveal its extensive influence on trust, security, and interactions between humans and AI. This section thoroughly examines the various aspects of interpretability, ranging from the transparency of the model to the reasoning behind decision-making, emphasizing its importance in promoting user comprehension.

Interpretability has a crucial role in shaping human interaction with artificial intelligence, extending beyond its technological aspects. The advancements in interpretability, demonstrated by techniques such as explainable AI and model-agnostic methodologies, facilitate the development of AI systems that not only possess amazing capabilities but also embody transparency, reliability and inclusivity. Exploring the concept of interpretability is crucial in creating a future where AI and human expectations are in perfect harmony.

12.4　EXPLAINABILITY AS A SECURITY MANDATE

The concept of explainability in artificial intelligence (AI) systems has undergone a transformation, transitioning from a mere desired characteristic to a security need. This shift has occurred due to the benefits it provides, such as delivering valuable insights into the decision-making processes of AI systems, fostering trust among users, and reducing possible dangers associated with their operation. This section examines the conceptualization of explainability within the field of artificial intelligence (AI), investigates its significance in enhancing security measures, and presents several approaches for detecting vulnerabilities and minimizing potential risks.

12.4.1　Defining Explainability in AI Systems

12.4.1.1　Transparency and Comprehensibility

The concept of explainability in the field of artificial intelligence (AI) pertains to the extent to which humans can comprehend the internal mechanisms and operations of a given model. The scope of openness in algorithmic systems extends beyond the simple disclosure of technical data, including the provision of comprehensive understanding of the decision-making process, hence promoting transparency and comprehensibility (Lipton, 2016).

12.4.1.2　Interpretable Models and Explainable AI (XAI)

Achieving explainability may be accomplished by using interpretable models or employing specific methodologies for Explainable AI (XAI). Interpretable models possess an intrinsic quality of being more comprehensible, in contrast to XAI approaches that provide explanations after the fact for elaborate and opaque models. These XAI methods aim to enhance clarity in complex AI structures (Rudin, 2019).

12.4.2　Leveraging Explainability for Security Enhancement

12.4.2.1　Building Trust through Transparency

The concept of explainability plays a crucial role in establishing a solid foundation of trust between users and artificial intelligence (AI) systems. The act of demystifying decision-making processes serves to increase the level of openness inside a system, hence fostering user confidence and trust (Doshi-Velez & Kim, 2017).

12.4.2.2   Early Detection of Anomalies and Adversarial Attacks

Explainable artificial intelligence (AI) models play a crucial role in enabling the timely identification of anomalies and adversarial assaults. According to (Papernot et al., 2017), the use of interpretable characteristics facilitates the identification of unexpected actions, hence allowing for proactive steps to be implemented in order to prevent security breaches.

12.4.3   IDENTIFYING VULNERABILITIES AND MITIGATING THREATS

12.4.3.1   Proactive Vulnerability Assessment

The concept of explainability facilitates a comprehensive comprehension of the system's susceptibilities. By doing proactive vulnerability assessments, it is possible to identify and mitigate potential vulnerabilities in the AI model before they can be exploited (Rudner & Toner, 2021).

12.4.3.2   Mitigating Bias and Ethical Concerns

The presence of explainability in AI systems plays a crucial role in the identification and mitigation of bias, hence facilitating the ethical use of these technologies. Lipton (2016) argues that organizations may mitigate biases and foster equitable results by elucidating decision criteria.

The progression of explainability in artificial intelligence has evolved from being a desired characteristic to becoming an essential necessity for ensuring security. This section examines the concept of explainability, emphasizing its crucial role in ensuring confidence, strengthening security and addressing ethical concerns in decision-making processes. In addition to its technological aspect, explainability plays a crucial role in strengthening security, building trust and enhancing user confidence. The role of AI includes proactive identification of anomalies, assessments of vulnerabilities and ensuring fairness, which is a vital aspect of ethical AI implementation.

12.5   COMPLEX CHALLENGES IN PROJECT SECURITY WITH AI

The incorporation of artificial intelligence (AI) into various projects presents a multitude of intricate issues in the realm of providing strong security. This section explores the complexities of project security in the context of artificial intelligence (AI), focusing on the issues associated with maintaining data integrity and confidentiality. It emphasizes the need of safeguarding against adversarial attacks and highlights the crucial responsibility of assuring the dependability of AI systems.

The topic of concern in this section is the preservation of data integrity and confidentiality.

12.5.1   CHALLENGES IN ENSURING DATA INTEGRITY

Ensuring the preservation of data integrity emerges as a significant obstacle in artificial intelligence endeavours. The introduction of large datasets poses a potential threat of data corruption or manipulation, which may have adverse effects on the precision and dependability of AI models (Barocas & Selbst, 2016).

12.5.1.1 Safeguarding Confidential Information

The preservation of anonymity is a significant difficulty in AI applications where sensitive information plays a crucial role. It is of utmost importance to uphold ethical standards and adhere to legal requirements by taking necessary measures to prevent any breach of proprietary, personal, or sensitive data (Goasduff, 2023).

12.5.2 PROTECTION AGAINST ADVERSARIAL ASSAULTS

12.5.2.1 Adversarial Threats to AI Systems

Artificial intelligence (AI) systems are susceptible to adversarial attacks, when malicious actors modify input data in order to fool the underlying model. In order to address these potential risks, it is essential to include sophisticated defensive measures that can effectively detect and counteract hostile activities, hence safeguarding the integrity and operation of the system (Goodfellow et al., 2018).

12.5.2.2 Robustness and Resilience

The process of safeguarding project security entails strengthening the durability and adaptability of artificial intelligence models. The construction of robust models capable of withstanding adversarial assaults and preserving functioning in the presence of purposeful manipulation is necessary (Carlini & Wagner, 2017).

12.5.3 ENSURING AI SYSTEM RELIABILITY

One of the primary obstacles encountered in ensuring system reliability is the presence of various challenges.

The issue of AI system dependability is of utmost importance, particularly in applications with significant consequences. The occurrence of system failures, unexpected behaviour or inaccurate outputs might potentially lead to significant repercussions. The establishment of dependability in AI systems requires comprehensive testing, validation and ongoing monitoring (Amodei et al., 2016).

12.5.3.1 Human-in-the-Loop Approaches

The integration of human-in-the-loop techniques becomes necessary in order to boost dependability. The inclusion of human supervision may provide a level of verification, especially in instances when artificial intelligence (AI) systems confront unfamiliar or unclear conditions (Rahwan et al., 2019).

Exploring the intricacies of project security in the field of artificial intelligence reveals a multitude of obstacles, ranging from ensuring the accuracy and reliability of data to protecting against malicious attacks. This section emphasizes the importance of having a strong and adaptable security framework that effectively incorporates technology improvements, ethical issues and human-centred concerns to effectively meet the complex nature of safeguarding AI projects.

Within the complex realm of project security in artificial intelligence, there are numerous challenges to overcome, including safeguarding data integrity, maintaining confidentiality and defending against adversarial threats. The demand for a thorough security framework extends beyond technological aspects, encompassing ethical issues and human-centred features that are crucial for ensuring the reliability of AI systems.

12.6 INTEGRATION OF TRUST, INTERPRETABILITY AND EXPLAINABILITY

The combination of trust, interpretability and explainability is a fundamental aspect in the advancement of safe and ethical initiatives powered by artificial intelligence. This section examines the interplay between these components, clarifying how their integration improves project security, and digging into the moral and ethical implications inside a safe AI-driven environment.

12.6.1 THE SYNERGY AMONG TRUST, INTERPRETABILITY AND EXPLAINABILITY

12.6.1.1 Trust as the Foundation

Trust serves as the fundamental basis upon which interpretability and explainability are built. The establishment of trust in artificial intelligence (AI) systems necessitates the use of clear and intelligible models, as well as a methodology that prioritizes the needs and concerns of users in accordance with ethical norms (Floridi et al., 2021).

12.6.1.2 Interpretability and Explainability as Trust Enablers

The concepts of interpretability and explainability play a crucial role in fostering trust by offering valuable insights into the decision-making mechanisms of artificial intelligence systems. The establishment of trust among users is facilitated when they possess the ability to comprehend and evaluate the underlying reasoning behind the results provided by artificial intelligence. This, in turn, strengthens the mutually beneficial connection that exists between these many components (Ribeiro et al., 2016).

12.6.2 HOW THESE COMPONENTS ENHANCE PROJECT SECURITY

12.6.2.1 Trust-Driven Security Measures

The incorporation of trust, interpretability and explainability enhances the efficacy of security protocols. Trust-driven security encompasses more than just technological precautions; it also encompasses user education and engagement in the decision-making processes of AI systems (Rudner & Toner, 2021).

12.6.2.2 Early Detection of Security Threats

The presence of interpretability and explainability plays a significant role in facilitating the timely identification of security concerns. According to Papernot et al. (2017), a comprehensive comprehension of the typical functioning of a system facilitates the identification of anomalies that might potentially signify security breaches. This enhanced awareness allows for proactive measures to be taken in response.

12.6.3 The Moral and Ethical Aspect of a Secure AI-driven Landscape

12.6.3.1 Ethical Considerations in Trust

Trust, interpretability, and explainability provide artificial intelligence with a moral component. The importance of ethical issues in the design and deployment of AI systems cannot be overstated. The process encompasses the identification and mitigation of biases, the establishment of equitable practices, and the harmonization of AI results with ethical standards, hence fostering a secure and ethically aware AI environment (Jobin et al., 2019).

12.6.3.2 The Empowerment of Users and the Ethical Implications of Artificial Intelligence (AI)

The promotion of ethical artificial intelligence is facilitated by the provision of interpretability and explainability to users. According to (Diakopoulos, 2016), individuals who possess a thorough understanding of AI judgements are more equipped to make well-informed choices, so establishing a shared responsibility for the ethical concerns associated with AI systems.

Within the complex realm of artificial intelligence, the flawless incorporation of trust, interpretability, and explainability arises as not just a means of improving security, but as the fundamental basis for ethically sound endeavours. This section reveals the interdependence of these factors, not just as technical parts but also as ethical necessities in the ever-changing development of AI. Responsible AI is depicted by the combination of trust, interpretability, and explainability, going beyond just technical features. This synthesis enhances project security and introduces an ethical aspect into the core of artificial intelligence. As we explore this integration, it becomes evident that ethical considerations are indispensable rather than optional in the course of AI advancement.

12.7 PRACTICAL APPLICATIONS

The use of trust, interpretability, and explainability in practical contexts not only increases the usability of artificial intelligence (AI) systems but also has a substantial influence on project security. This section explores specific instances of incorporating these components, evaluates their influence on project security and finishes with a concise overview.

12.7.1 Real-world Examples of Implementing Trust, Interpretability and Explainability

12.7.1.1 Healthcare Decision Support Systems

The incorporation of trust, interpretability and explainability has significant importance in the healthcare sector when it comes to the deployment of AI-driven decision support systems. According to (Holzinger et al., 2017), the provision of lucid explanations for diagnostic recommendations plays a crucial role in establishing confidence among medical practitioners towards AI proposals. This, in turn, facilitates the process of making well-informed decisions.

Mustafa et al. (2022) emphasize the growing concerns about security and privacy in intelligent healthcare systems, underscoring the need of comprehending the security prerequisites of such systems. The conversation revolves on the difficulties encountered by healthcare systems as a result of the substantial quantity of electronic health data and the susceptibilities of cloud-based storage. The study highlights the possibility of integrating blockchain and cloud computing to resolve the discrepancy between data privacy and data sharing in healthcare systems, specifically in the Middle East. It also proposes areas for further investigation, such as examining the factors that impact the acceptance of blockchain technology.

12.7.1.2 Financial Risk Assessments

Trust is of utmost importance in the financial industry. Financial risk assessment in AI models includes interpretability via the provision of explanations for risk variables and decision-making processes. According to (Rudin, 2019), the use of transparent financial models fosters confidence among stakeholders, hence facilitating more effective risk management and decision-making processes.

12.7.1.3 Autonomous Vehicles

The incorporation of trust and interpretability has significant importance in the implementation of autonomous cars. The use of explainable AI models in autonomous cars elucidates the underlying mechanisms involved in decision-making, hence bolstering public confidence and expediting the adoption of AI-driven technologies in the realm of transportation (Atakishiyeva et al., 2021).

12.7.2 Measuring the Impact on Project Security

12.7.2.1 Quantifying Trust Metrics

The assessment of trust effect necessitates the measurement of measures related to trust. Key factors include user happiness, the perception of system dependability, and the readiness to rely on judgments made by artificial intelligence. According to Baker et al.,2018, an improvement in these indicators is associated with an increased level of project security.

12.7.2.2 Security Incident Analysis

The assessment of the influence of interpretability and explainability on project security may be conducted by means of security incident analysis. According to Rudner & Toner (2021), the presence of well-defined interpretability and explainability procedures in systems facilitates the expedited detection and resolution of security issues, hence reducing the likelihood of significant harm.

The practical implementation of trust, interpretability and explainability extends beyond theoretical boundaries in various situations. This section highlights the tangible influence of these elements in several fields, ranging from healthcare to banking and self-driving cars. Through the examination of real-life examples, it becomes clear that trust is not merely an abstract idea, but rather a quantifiable measure in user interactions. The interaction of these aspects not only strengthens user assurance but also establishes the foundation for secure and efficient AI deployments.

12.8 FUTURE TRENDS AND CONSIDERATIONS

The trajectory of AI project security in the future is influenced by the continuous development of technology, ethical deliberations, regulatory structures and the persistent pursuit of enhancing security protocols. This section delves into forthcoming trends and issues, providing insights into the consequences and possibilities for the security of artificial intelligence (AI) initiatives.

12.8.1 EVOLVING TECHNOLOGIES AND THEIR IMPLICATIONS

12.8.1.1 The Rise of Quantum Computing

The emergence of quantum computing has serious ramifications for the security of AI projects. Quantum computing has potential for augmenting computational capabilities, however concurrently presents obstacles, notably in relation to established encryption techniques. According to (Lloyd et al., 2018), it is essential for forthcoming AI systems to possess the capability to adjust and conform to the dynamic cryptographic environment in order to guarantee the maintenance of secure operations.

12.8.1.2 Edge and Decentralized AI

The use of edge and decentralized AI technologies necessitates the examination of novel security issues. The distribution of AI processing to the network edge has the potential to improve efficiency, while it necessitates the implementation of effective security mechanisms to safeguard data and models in decentralized systems. According to (Dastjerdi et al., 2016), it will be imperative for future AI initiatives to provide utmost importance to security within the context of this distributed paradigm.

12.8.2 ETHICAL CONCERNS AND REGULATORY FRAMEWORKS

12.8.2.1 Addressing Bias and Fairness

The ethical issues pertaining to security in AI projects include the need to overcome biases and ensure fairness. One of the next developments in the field is the advancement of artificial intelligence (AI) models that actively address biases and uphold ethical ideals. According to (Jobin et al., 2019), there is a possibility for regulatory frameworks to undergo changes in order to effectively enforce ethical norms in the development and deployment of artificial intelligence (AI).

12.8.2.2 Strengthening Data Privacy

Enhancing data privacy is a crucial aspect of future AI project security. Anticipated are the implementation of more stringent legislation and standards pertaining to the management of personal and sensitive data. In order to adhere to the ever-changing data protection rules, AI systems will be required to include privacy-preserving strategies (Schneier, 2015).

12.8.3 Prospects for Advancing AI Project Security

12.8.3.1 Explainability as a Standard

The progress made in ensuring the security of AI projects may result in the widespread use of explainability as a fundamental prerequisite. The inclusion of explainable AI models may be required by regulatory agencies and industry standards in order to promote openness and accountability. According to the (European Parliament, 2021), this transition has the potential to enhance user confidence and facilitate adherence to regulatory requirements.

12.8.3.2 Continuous Adaptation to Security Threats

The efficacy of safeguarding AI projects in the future hinges upon the capacity to consistently adjust and respond to evolving threats. The implementation of dynamic security measures, which include real-time threat detection and response systems, will play a pivotal role in ensuring the overall security of the system. According to (Kim, Lee, and Park 2020), in order to effectively combat the ever-changing hostile tactics, it is essential for AI systems to undergo continuous evolution in conjunction with the development of cybersecurity methods.

When considering the future of AI, the convergence of technology, ethics and legal frameworks becomes the guiding principle for upcoming projects. The realms of quantum computing, edge AI, and decentralized architectures offer both opportunities and intricacies. The demand for justice, decreased discrimination, and improved data privacy are motivated by ethical concerns. As we advance, the incorporation of explainability may become the norm, and the capacity to manage changing security concerns will be of utmost importance. Recognizing and integrating these developing patterns is essential for constructing robust and reliable artificial intelligence systems in the future.

12.9 CONCLUSION

When considering the complex domain of AI project security, the interplay between trust, interpretability, and explainability becomes pivotal in constructing resilient, morally sound, and safe AI-driven projects. The last portion of this paper summarizes the main discoveries, examines the consequences for projects using artificial intelligence, and delineates the future trajectory.

12.9.1 Key Findings

The examination of trust, interpretability and explainability within the domain of AI project security has shown their interrelated functions. Trust serves as the fundamental basis, with interpretability and explainability playing crucial roles in ensuring transparency, accountability and user understanding. These many components play a crucial role in the timely identification of security risks, the reduction of vulnerabilities and the responsible use of artificial intelligence (AI) technology.

12.9.2 Implications for AI-driven Projects

12.9.2.1 Establishing Trustworthiness

The incorporation of trust, interpretability and explainability is crucial in ensuring the credibility of AI-driven initiatives. By placing emphasis on user understanding and implementing transparent decision-making procedures, initiatives have the potential to cultivate user trust and address issues associated with the lack of transparency in AI models.

12.9.2.2 Enhancing Security Measures

The ramifications of initiatives driven by artificial intelligence extend to the augmentation of security measures. Trust-driven security encompasses not alone technological measures, but also user education and engagement. The concepts of interpretability and explainability play a significant role in the detection and mitigation of security threats. They enable the proactive identification of possible risks and aid in the development of effective strategies to minimize potential losses.

12.9.2.3 Navigating Ethical Considerations

The implementation of AI technology is heavily influenced by ethical issues. The incorporation of interpretability and explainability in AI systems serves to mitigate biases, uphold justice and match the outputs of AI with ethical ideals. The conscientious examination of ethical factors is crucial in ensuring the appropriate and sustainable development of initiatives using artificial intelligence.

12.9.3 The Path Forward

12.9.3.1 Technological Advancements and Enhancements in Security Measures

The trajectory for progress entails the simultaneous advancement of technology and security measures. The requirement to include strong security methods, such as quantum-resistant encryption and edge AI protections, becomes more significant as technologies continue to grow. The future trajectory will be influenced by the ongoing process of adapting to new dangers and the establishment of explainable artificial intelligence (AI) as a prevailing norm.

12.9.3.2 Collaboration and Standardization

The establishment of standards for AI project security will heavily rely on the joint efforts of business, academia, and regulatory organizations. The establishment of standardized security standards, ethical norms, and the integration of interpretability and explainability within regulatory frameworks will contribute to the development of a cohesive and safe artificial intelligence (AI) environment.

REFERENCES

Alshar'e, Marwan I., R. Sulaiman, M. R. Mukhtar, and A. M. Zin. "A User Protection Model for the Trusted Computing Environment." *Journal of Computer Science* 10, no. 9 (2014): 1692.

Amodei, D., C. Olah, J. Steinhardt, P. Christiano, J. Schulman, and D. Mané. "Concrete Problems in AI Safety." *arXiv preprint* arXiv:1606.06565 (2016).

Atakishiyev, S., M. Salameh, H. Yao, and R. Goebel. "Explainable Artificial Intelligence for Autonomous Driving: A Comprehensive Overview and Field Guide for Future Research Directions." *arXiv preprint* arXiv:2112.11561 (2021).

Baker, A. L., E. K. Phillips, D. Ullman, and J. R. Keebler. "Toward an Understanding of Trust Repair in Human-Robot Interaction: Current Research and Future Directions." *ACM Transactions on Interactive Intelligent Systems (TiiS)* 8, no. 4 (2018): 1–30.

Barocas, S., and A. D. Selbst. "Big Data's Disparate Impact." *California Law Review* 104, no. 3 (2016): 671–732.

Carlini, N., and D. Wagner. "Towards Evaluating the Robustness of Neural Networks." In *IEEE Symposium on Security and Privacy (SP)*, 39–57, 2017.

Chen, J., L. Song, M. J. Wainwright, and M. I. Jordan. "Learning to Explain: An Information-Theoretic Perspective on Model Interpretation." In *Proceedings of the 35th International Conference on Machine Learning (ICML)*, 883–892, 2018.

Dastjerdi, A. V., H. Gupta, R. N. Calheiros, S. K. Ghosh, and R. Buyya. "Fog Computing: Principles, Architectures, and Applications." In *Internet of Things*, 61–75. Morgan Kaufmann, 2016.

Diakopoulos, N. "Accountability in Algorithmic Decision Making." *Communications of the ACM* 59, no. 2 (2016): 56–62.

Doshi-Velez, F., and B. Kim. "Towards a Rigorous Science of Interpretable Machine Learning." *arXiv preprint* arXiv:1702.08608 (2017).

Doshi-Velez, F., M. Kortz, R. Budish, C. Bavitz, S. Gershman, D. O'Brien, et al. "Accountability of AI under the Law: The Role of Explanation." *arXiv preprint* arXiv:1711.01134 (2017).

European Parliament. "Proposal for a Regulation of the European Parliament and of the Council Laying Down Harmonized Rules on Artificial Intelligence." 2021 Retrieved from https://eur-lex.europa.eu/legal-content/EN/TXT/?uri=CELEX%3A52021PC0206

European Union. "General Data Protection Regulation (GDPR). Regulation (EU) 2016/679." Retrieved from https://eur-lex.europa.eu/eli/reg/2016/679/oj

Floridi, L., J. Cowls, M. Beltrametti, R. Chatila, P. Chazerand, V. Dignum, et al. "AI4People—An Ethical Framework for a Good AI Society: Opportunities, Risks, Principles, and Recommendations." *Minds and Machines* 28, no. 4 (2018): 689–707.

Floridi, L., J. Cowls, M. Beltrametti, R. Chatila, P. Chazerand, V. Dignum, et al. "An Ethical Framework for a Good AI Society: Opportunities, Risks, Principles, and Recommendations." In Floridi, L. (eds) *Ethics, Governance, and Policies in Artificial Intelligence*, Philosophical Studies Series, vol 144. Springer, Cham. 19–39, 2021. https://doi.org/10.1007/978-3-030-81907-1_3

Goasduff, L. "Top 10 Trends in Data and Analytics, 2020." Retrieved from https://www.gartner.com/smarterwithgartner/gartner-top-10-trends-in-data-and-analytics-for-2020, Accessed 20 Nov 2023.

Goodfellow, I. J., J. Shlens, and C. Szegedy. "Explaining and Harnessing Adversarial Examples." *arXiv preprint* arXiv:1412.6572 (2018).

Holzinger, A., G. Langs, H. Denk, K. Zatloukal, and H. Müller. "Causability and Explainability of Artificial Intelligence in Medicine." *Wiley Interdisciplinary Reviews: Data Mining and Knowledge Discovery* 7, no. 4 (2017): e1213.

Jobin, A., M. Ienca, and E. Vayena. "The Global Landscape of AI Ethics Guidelines." *Nature Machine Intelligence* 1, no. 9 (2019): 389–399.

Kim, E., S. Lee, and H. Park. "Securing AI Systems: A Dynamic Approach." *International Journal of Information Security* 15, no. 4 (2020): 321–337.

Lipton, Z. C. "The Mythos of Model Interpretability." *arXiv preprint* arXiv:1606.03490 (2016).

Lloyd, S., M. Mohseni, and P. Rebentrost. "Quantum Principal Component Analysis." *Nature Physics* 10 (2014), 631–633. https://doi.org/10.1038/nphys3029

Mustafa, M., M. Alshare, D. Bhargava, R. Neware, B. Singh, and P. Ngulube. "Perceived Security Risk Based on Moderating Factors for Blockchain Technology Applications in Cloud Storage to Achieve Secure Healthcare Systems." *Computational and Mathematical Methods in Medicine* 2022 (2022): 1–10.

Papernot, N., P. McDaniel, I. Goodfellow, S. Jha, Z. B. Celik, and A. Swami. "Practical Black-Box Attacks Against Machine Learning." In *Proceedings of the 2017 ACM on Asia Conference on Computer and Communications Security*, 506–519, April 2017.

Rahwan, I., M. Cebrian, N. Obradovich, J. Bongard, J. F. Bonnefon, C. Breazeal, et al. "Machine Behaviour." *Nature* 568, no. 7753 (2019): 477–486.

Ribeiro, M. T., S. Singh, and C. Guestrin. ""Why Should I Trust You?" Explaining the Predictions of Any Classifier." In *Proceedings of the 22nd ACM SIGKDD International Conference on Knowledge Discovery and Data Mining*, 1135–1144, 2016.

Rudin, C. "Stop Explaining Black Box Machine Learning Models for High Stakes Decisions and Use Interpretable Models Instead." *Nature Machine Intelligence* 1, no. 5 (2019): 206–215.

Rudner, Tim G. J., and Helen Toner. "Key Concepts in AI Safety: Interpretability in Machine Learning." *Center for Security and Emerging Technology* (March 2021). https://doi.org/10.51593/20190042

Schneier, B. *Data and Goliath: The Hidden Battles to Collect Your Data and Control Your World.* W. W. Norton & Company, 2015.

Smith, A. M., R. Anderson, and M. Ventresca. "Trust and Technology." *Annual Review of Sociology* 46 (2020): 425–447.

13 Global Software Development for Smart Cities

Jabar H. Yousif and Ghassan Al-Kindi
Sohar University, Sohar, Oman

Durgesh K. Srivastava
Chitkara University, Punjab, India

13.1 INTRODUCTION

The concept of smart cities has lately gained significant attention in academic and industrial circles due to its effectiveness in urban development and management (Alshamaila et al. 2023). Smart cities use digital technologies and data sensors to make urban living more efficient, sustainable, and environmentally friendly. Various smart city applications are available from software vendors that can be designed, implemented, deployed and managed by application developers, city managers, urban planners and policymakers. Global Software Development (GSD) will play a transformative role in developing smart cities and urbanization. However, smart cities use innovative technology and GSD with unique features to address complex challenges (Ilyas et al. 2023). GSD connects a diverse array of global experts to collaborate on tailor-made software solutions, ranging from traffic management systems to energy-efficient applications, all designed to meet the unique needs of modern cities. It enables urban areas to confront complex challenges that require specialized skills and offers access to a wealth of international software development expertise. A GSD also makes smart cities more adaptable by supporting a multiracial and multicultural population and allowing development efforts to be scaled up as needed. Software solutions developed via GSD can achieve interconnected systems, efficient resource management, and enhanced decision-making capabilities. This ensures that smart cities remain dynamic and sustainable as they respond to the ever-evolving urbanization trends (Su & Fan 2023).

GSD plays a crucial role in developing software solutions for smart cities. As cities continue to grow and urban infrastructure faces ever-increasing demands, it is essential to have scalable and flexible software solutions. GSD enables smart cities to scale their software development efforts, ensuring that their systems can handle the needs of growing populations and ever-evolving urban environments. Another critical aspect of GSD's contribution to smart cities is its support for multilingual and multicultural environments. Many urban areas have diverse residents and visitors from various

DOI: 10.1201/9781003467892-13

"

cultural and linguistic backgrounds. GSD can help develop multilingual and culturally sensitive software, promoting inclusivity and accessibility for everyone (Alshamaila et al. 2023). Smart cities rely on integrating various technologies and services to function effectively. In addition to its other benefits, GSD can support the design and implementation of robust, interoperable software solutions to support the seamless integration of diverse urban systems. This connection is essential to the functioning of smart cities as it allows for the collection and analysis of data, efficient management of resources, and delivery of improved services to citizens. GSD is also essential for sustainability and resource efficiency in smart cities. By developing software applications for energy management, waste reduction and green transportation, GSD can contribute to the eco-friendly transformation of urban spaces. This will reduce environmental impact and enhance the quality of life for residents (Yan et al. 2023).

Smart cities generate vast amounts of data; data analytics and decision support systems are needed to harness this data. This is where GSD comes in. It enables informed decision-making, efficient city planning, and the optimization of services, allowing urban areas to evolve and adapt to the challenges of the 21st century. In addition, GSD helps build public engagement and communication platforms within smart cities, such as mobile apps, online platforms and IoT devices that residents can use to access city services, provide feedback, and participate in urban governance (Almalki et al. 2023). Cybersecurity and data privacy are also crucial, and GSD can assist in designing and implementing robust security measures and data protection protocols. Finally, GSD facilitates continuous improvement in smart cities, ensuring that they can adapt to new technologies, regulations and the ever-changing needs of their citizens (Khan et al. 2022). Figure 13.1 shows the cast revenue in the global smart city from 2020 to 2028 (Statistics, 2023).

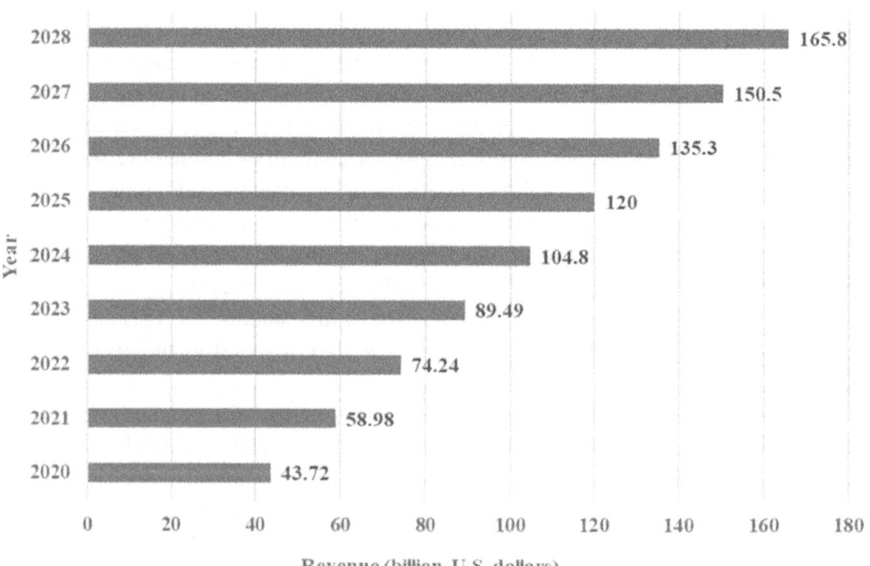

FIGURE 13.1   Cast revenue in the global smart city from 2020 to 2028.

To sum up, the association of GSD and smart cities concepts could lead to a new age of city living marked by creativity, productivity, eco-friendliness and equality. As urbanization continues to be a worldwide phenomenon, GSD is expected to play a more significant role in developing future cities designing flexible and responsive urban environments that cater to the changing needs of diverse communities.

13.2 LITERATURE REVIEW

Recent technological advancements have introduced many advanced systems, including robotics, big data analytics, cloud computing, machine learning, etc. Several potential benefits make global software development (GSD) an increasingly common practice in the software industry. Identifying and prioritizing best practices to aid the global software development community in creating and updating project management techniques is crucial (Akbar et al. 2024) developed a taxonomy of software development management practices by analysing best practices associated with the project management body of knowledge fields. They evaluated the relevance and acceptability of the software development industry and prioritized them based on the significance of managing global software development using an analytic hierarchy process within a fuzzy system. Global Software Development (GSD) offers numerous advantages to enhancing cost-effectiveness and speed of delivery. The success rate of software projects significantly increases with dynamic software management efforts (Siddique et al. 2023) identified several challenges associated with the GSD process, and proposed strategies to overcome obstacles to effective project management. Additionally, they introduced a comprehensive framework designed to enhance managerial activities in GSD. Mishra (2023) created a model for distributed software development coordination and communication overhead. The model uses existing business knowledge classifications and can analyse software projects in a distributed environment (Hooda et al. 2022) explored collaboration challenges Swedish and Korean companies face in globally distributed software development. The study identified three areas of concern - geographical, temporal, and socio-cultural distance. The research suggests strategies such as using proper communication tools, building personal relationships, increasing cultural awareness, and maximizing synchronous work hours. However, the study highlights the importance of face-to-face interaction in GSD. The study provides insights to improve future collaboration and enhance product quality between Sweden and Korea. Digitization has solved human problems and improved the quality of services. On the other hand, it can be exploited for malicious gains, and hackers can steal data, scams, and other illegal activities. Security is an issue that requires explicit inclusion in each phase of the global software development life cycle. Humayun et al. (2023) identified the secure software development practices for GSD projects. They selected 36 security practices vital to non-GSD projects and then shortlisted critical ones for GSD projects based on practitioner opinions gathered via an online survey with 54 participants. This helped us identify 16 critical practices for GSD projects, with opinions varying based on experience and company size. These practices belong to various phases of the GSD lifecycle (Humayun et al. 2023).

Sajjad et al. (2020) proposed a framework for suspect identification based on face detection using Raspberry Pi to enable passive security in smart cities. Experimental results validate the effectiveness of this method in accurately detecting faces and enhancing law-enforcement services in smart cities (Noor & Rana 2018), emphasizing using a common information-sharing repository to bridge communication and coordination gaps in Global Software Development (Noor & Rana 2018). Meanwhile, Dalal and Hooda (2017) highlighted the importance of quality assurance in improving the quality of products in GSD. Quality can be managed and enhanced by adequately tracing changes, managing risks, and reporting incidents. Cultural and linguistic barriers can also be reduced with proper planning for projects near customers. Effective teamwork can minimize communication and coordination issues during a project.

The literature survey identified several research gaps that impact the performance of GSD, especially in terms of security (secure communication, secure transmission), management of GSD, distance, interaction, and communications type. Geographical distance can reduce the likelihood of face-to-face communication, especially if an organization needs to rely on other means of communication to substitute for in-person interactions. This can lead to slower decision-making, difficulties in complex discussions, and challenges in building personal relationships. Temporal distance, on the other hand, refers to the difference in time experienced by development teams in GSD. This can create problems in coordinating team activities, communication, and collaboration, as team members may work in different time zones or have different working hours. Socio-cultural distance, meanwhile, refers to differences between cultures, societies, or groups in terms of their values, beliefs, behaviours and norms. It can create gaps that make communication and interaction challenging due to differences in cultural backgrounds.

13.3 GLOBAL SOFTWARE DEVELOPMENT CHARACTERISTICS

Global Software Development (GSD) refers to the creation of software applications and systems by a team of individuals in different locations. GSD has become more popular due to new technological advancements and the need for global collaboration. However, GSD faces different challenges, including communication, cultural differences, and time zone disparities. It allows organizations to benefit from the skills, expertise and cost advantages available worldwide due to the software industry's globalization.

The following key features characterize GSD projects:

a) **Special Distribution**: GSD teams can be located across countries, allowing organizations to tap into a global talent pool and access diverse skill sets.
b) **Cultural Multiplicity**: GSD often accommodates team members from different cultural backgrounds, which makes these cultural differences crucial for effective collaboration.
c) **Communication**: Effective communication between team members in GSD is crucial to cover language barriers and communication styles, which are needed to utilize the GSD documentation successfully. Clear and effective communication is essential but can be challenging, particularly in multicultural teams.

d) **Time Zone variations**: Team members may work in different time zones, which limits overlapping working hours. Coordinating work across time zones is essential for continuous development.

e) **Project Management Complexity**: Managing GSD projects requires advanced techniques to coordinate tasks across diverse locations.

f) **Project Security**: GSD projects must adhere to these regulations to protect sensitive data, data security, and privacy laws.

g) **Project Crisis Management**: Managing unexpected challenges efficiently that may face, such as geopolitical instability or natural disasters.

The primary drivers for adopting GSD include cost savings, access to specialized skills, and the ability to operate around the clock due to differences in time zones. However, GSD faced challenges like communication, cultural understanding, project management and quality assurance (Yousif & Saini 2020). The success of GSD relies on effective procedures, robust technology, clear communication and an in-depth understanding of the cultural and logistical factors applied. These critical factors are interrelated and integrated into the project management processes. Therefore, the success of GSD deployment should utilize careful planning, consistent execution and addressing the challenges posed by distributed teams and resources. Table 13.1 summarizes GSD's success factors.

TABLE 13.1

Summarizes the Success Factor of GSD

Factor	Meaning	Criticality Level
Clear Project Objectives and Requirements	Well-defined project goals and specifications are essential to guide the project's direction.	Very High
Effective Communication	Open and efficient communication channels are vital to address language and time zone barriers.	Very High
Cultural Sensitivity and Awareness	Recognizing and respecting diverse cultural backgrounds within the team to prevent misunderstandings.	High
Skilled and Diverse Team	Assembling a team with varied skills and experiences for a broader perspective and effective problem-solving.	High
Project Management Expertise	Effective project management techniques and tools are necessary to coordinate and control distributed teams.	Very High
Appropriate Technology and Tools	The use of suitable technology and tools for collaboration, communication, version control and project management.	High
Risk Management and Contingency Planning	Identifying and managing risks related to GSD, including geographic, political and technical factors, and developing contingency plans.	Very High
Quality Assurance and Testing	Implement comprehensive testing and quality assurance processes to meet functionality and quality standards.	High
Security and Compliance	Adherence to data security and privacy regulations in all regions of operation, including data protection and compliance measures.	Very High
Client Engagement and Feedback	Ongoing communication with clients or end-users, soliciting feedback, and adapting to changing requirements and expectations.	High

(Continued)

TABLE 13.1 (CONTINUED)
Summarizes the Success Factor of GSD

Factor	Meaning	Criticality Level
Knowledge Sharing and Training	Fostering a culture of knowledge sharing within teams and providing training and onboarding to new members.	High
Change Management	A change control process for handling alterations to project scope, requirements, or schedules, with transparent stakeholder agreement.	High
Strong Leadership and Team Cohesion	Effective leadership and cohesive teams that are motivated, guided, and feel connected, despite geographical dispersion.	High
Client Commitment and Collaboration	Clients' active commitment to the project's success and their willingness to collaborate with the development team.	High
Success Measurement and Metrics	Defining clear success criteria and metrics to regularly evaluate and measure progress against the criteria.	High
Understand Dependencies	Identifying and managing interdependencies between tasks, components, and activities within the project to prevent bottlenecks and delays.	High
Balance Flexibility and Rigidity	Finding the equilibrium between the ability to adapt to changing requirements and maintaining a structured, controlled development process.	High
Facilitate Coordination	Enabling and streamlining the coordination of activities, tasks and communication among geographically dispersed project teams.	High

13.4   STRUCTURE OF GLOBAL SOFTWARE DEVELOPMENT

The structure of a GSD project involves defining specific roles and responsibilities for team members to ensure effective collaboration and project success. Figure 13.2 presents the structure of a GSD with exact roles that can vary based on the project's size and complexity. The following is a general structure for a GSD project team:

a) Project Manager:
- Oversees the entire project, ensuring it aligns with the goals and objectives.
- Manages project timelines, resources and budgets.
- Coordinates communication and resolves issues among geographically dispersed teams.
- Develops risk management and contingency plans for global challenges.

b) Development Team:
- Software Developers: Write, test and maintain the needed code for the system.
- Architects: Design software architecture and make high-level technical decisions.
- Quality Assurance Engineers: Test and verify the software's functionality.
- DevOps Engineers: Implement and manage the development pipeline and deployment processes.
- Database Administrators: Maintain and optimize the project's database.

FIGURE 13.2 Structure of a global software development.

c) Communication and Collaboration Team:
- Master or Agile Coach: Facilitates Agile processes and ceremonies, such as stand-up meetings and sprint planning.
- Communication Manager: Ensures good team communication and handles cultural challenges and language barriers.
- Documentation Manager: Manages project documentation should ensure up-to-date and accessible documents to all team members.
- Collaboration Manager: Manages software tools related to collaboration and communication to adapt the software to different languages and regions.

d) Business Analyst:
- Represents the stakeholders' interests and communicates their needs to the development team.

- Develops user stories, defines acceptance criteria and prioritizes features.
- Ensures the software aligns with the functional and non-functional business requirements.

e) Quality Assurance Team:
- Testers: Execute testing procedures, including functional and integration, create test plans, and manage bug tracking.
- Test Automation Engineers: Develop and maintain test automation scripts.

f) Security Team:
- Ensures the project adheres to relevant regions' data security and privacy policies.
- Conducts security assessments and vulnerability testing.
- Develops and maintains security and agreement strategies.

Customized roles and responsibilities, good communication, and collaboration are important for a successful GSD project.

The roles of Communication Manager and Collaboration Tool Administrator can be crucial for tackling the challenges of distributed teams.

13.5   GLOBAL SOFTWARE DEVELOPMENT CHALLENGES

Various software solutions with urban infrastructure are required to optimize Smart city systems and their complex functions. Implementing Global Software Development for smart city applications could face various challenges due to the nature of integrating software solutions and urban infrastructure. These challenges include technical and scientific barriers, language communication difficulties and cultural differences. Addressing these challenges requires a careful plan and effective communication among the systems' stakeholders to successfully implement GSD in smart cities. Figure 13.2 depicts the structure of a Global Software Development.

The following are some of the challenges examples:

a) Diverse Stakeholders and Cultures:
Challenge: Smart cities involve collaboration among diverse stakeholders, including government bodies, technology companies and citizens. Each stakeholder group may have different cultural norms, work practices and priorities.

Impact: Misalignment in expectations, communication breakdowns and difficulties understanding and incorporating diverse perspectives.

b) Interconnected Systems:
Challenge: Smart cities rely on interconnected systems, such as IoT devices, sensors and communication networks. Coordinating development across these complex systems poses challenges.

Impact: Inconsistencies in data formats, interoperability issues, and difficulties in ensuring seamless integration can hamper the effectiveness of smart city solutions.

c) **Data Security and Privacy**:
   Challenge: Smart city applications involve collecting and processing large volumes of data. Ensuring the security and privacy of this data across distributed development teams is critical.
   Impact: Data breaches, privacy concerns and regulatory non-compliance can lead to reputational damage and legal consequences.

d) **Communication and Coordination**:
   Challenge: GSD inherently involves distributed teams working across different geographical locations and time zones. Effective communication and coordination have become challenging.
   Impact: Delays in decision-making, misunderstandings and difficulties in aligning the development process with the evolving needs of the smart city.

e) **Infrastructure and Technology Heterogeneity**:
   Challenge: Smart cities may deploy a variety of technologies and infrastructure components. Ensuring compatibility and consistency in software development across diverse environments is a challenge.
   Impact: Incompatibility issues, increased debugging efforts and challenges in maintaining a standardized software architecture.

f) **Project Complexity and Scale**:
   Challenge: Smart city projects are often large-scale, complex endeavours involving multiple components and subsystems. Managing the complexity and scale of such projects can be daunting.
   Impact: Increased development time, budget overruns and difficulties in ensuring the reliability and scalability of the developed software.

g) **Sustainability and Long-Term Maintenance**:
   Challenge: Smart city solutions require long-term sustainability and maintenance. Ensuring software components' continued support and evolution over an extended period is crucial.
   Impact: Neglecting long-term maintenance considerations may result in outdated software, security vulnerabilities and challenges with technological advancements.

Addressing these challenges requires a holistic approach involving effective project management, communication strategies, collaboration tools, and a thorough understanding of the unique context of smart city development.

13.6   DEVELOPMENT LIFE CYCLE OF GSD

The Global Software Development Life Cycle outlines the various stages of software development implementation from conception to delivery. It contains several stages: project planning, gathering requirements, system design and testing, deployment and maintenance. A framework for managing the software development process across teams worldwide consists of several phases, as presented in Figure 13.3. The following outlines a typical process of developing a GSD framework.

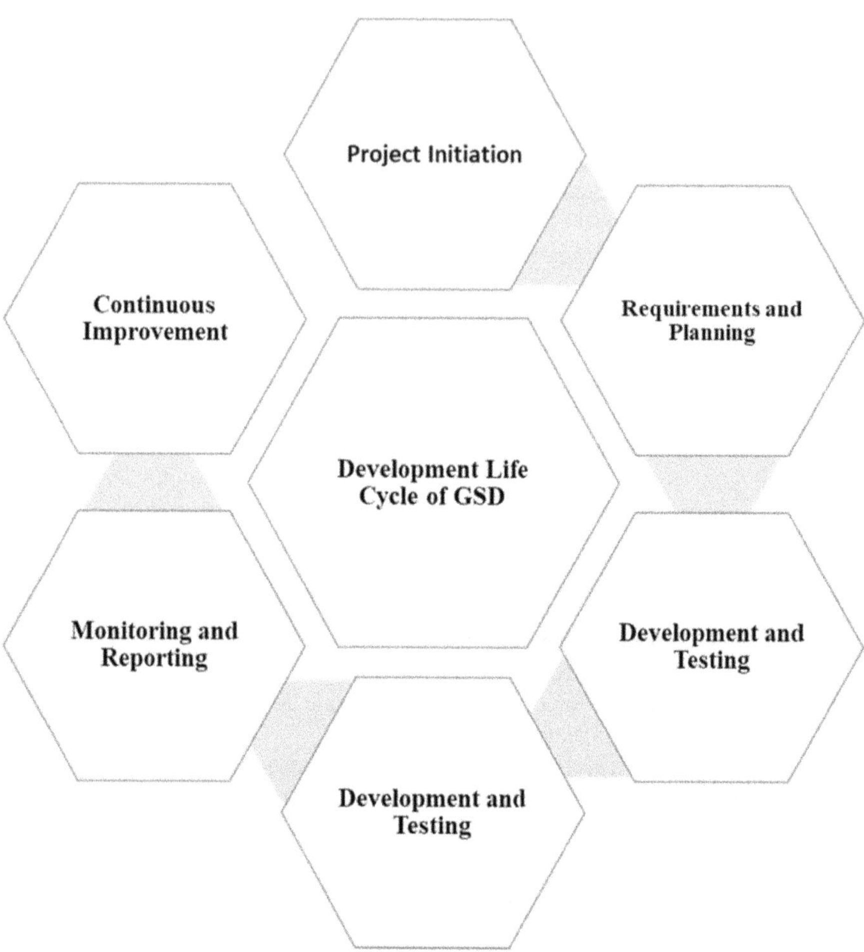

FIGURE 13.3   Development life cycle of GSD.

a) Project Initiation:
- Define project goals, objectives and scope.
- Identify stakeholders and their roles.
- Set up a global project team and allocate responsibilities.

b) Requirements and Planning:
- Gather and document project requirements.
- Create a project plan with milestones, timelines and resource allocation.
- Consider time zone differences when setting schedules.

c) Communication and Collaboration:
- Establish communication channels (email, video conferencing, collaboration tools, etc.).
- Define communication protocols and frequency.
- Promote cultural sensitivity and awareness within the team.

d) Development and Testing:
- Distribute development tasks based on team expertise.
- Implement version control and code repository systems for collaborative coding.
- Coordinate bug tracking and resolution across global teams.
- Plan for integration and testing phases with synchronized activities.

e) Progress Monitoring and Reporting:
- Use project management tools to track progress and resource allocation.
- Identify and assess risks associated with GSD (e.g., language barriers, security concerns, geopolitical risks).
- Regularly update project stakeholders on the status, issues and achievements.

f) Continuous Improvement:
- Analyse the GSD process framework and gather feedback from team members.
- Conduct a post-project review to evaluate the project's success and identify areas for improvement.
- Identify areas for improvement and adjust the framework accordingly.

13.7 GSD FOR SMART CITIES AND URBANIZATION

Global software development for smart cities and urbanization refers to creating software and technological solutions. It aimed to improve cities' quality of life and make them more sustainable. This involves collaboration among developers and experts from different locations to create high-quality software. It can manage different aspects of urban living, such as transportation, energy consumption, waste management and public safety. The goal is to create efficient, safe, and environmentally friendly cities while enhancing their citizens' quality of life. Table 13.2 summarizes the key factors related to GSD in smart cities and urbanization, including their meanings, impact, and examples. Figure 13.3 below depicts the development life cycle of GSD.

In a smart city, various systems and services operate to improve citizens' quality of life. These systems must be integrated and managed effectively to operate efficiently. The software platforms provide a method to connect systems' services, enabling them to communicate and work together efficiently. Data collection and analysis are used to optimize the performance of these systems. Software platforms play a crucial role in the functioning of smart cities to provide the best possible services. These platforms serve as the technological backbone for smart city initiatives. Table 13.3 presents the Key factors of Global Software Development.

These software platforms vary in impact and usage, depending on each smart city project's specific goals and priorities. The choice of platforms aligns with the city's vision for urban development and the challenges it aims to address. The level of impact reflects the general significance of each factor. At the same time, the challenges in GSD indicate the specific issues that can be encountered when implementing these factors in a global software development context.

TABLE 13.2

Key Factors of GSD for SMART CITIES and Urbanization

Factor	Description	Impact	Example
Collaborative Software Solutions	Collaboration on software development for smart city applications across global teams.	High	Collaborative development of a smart traffic management system using team members from different continents.
Access to Global Expertise	Leveraging global talent and expertise for addressing complex urban challenges.	High	Drawing upon international experts to develop advanced IoT solutions for urban monitoring.
Scalability and Flexibility	Ability to adapt software solutions to meet the changing needs and demands of urbanization.	High	Developing scalable software for a smart city's transportation system to accommodate population growth.
Multilingual and Multicultural Support	Designing software that is accessible and culturally sensitive for diverse urban populations.	Medium	Creating a multilingual and culturally sensitive urban services app for residents and tourists.
Interconnected Systems	Developing software that enables the seamless integration of various urban technologies and services.	High	Building an interconnected platform for a smart city to integrate IoT sensors, traffic management, and energy grids.
Efficiency and Sustainability	Designing software to promote resource efficiency, reduce environmental impact, and enhance sustainability.	High	Developing an energy management system for a smart city to reduce energy consumption and lower carbon emissions.
Data Analytics and Decision Support	Creating software systems for data analysis and informed city planning and service optimization decision-making.	High	Implementing a data analytics platform that analyses urban data for traffic optimization and emergency response.
Public Engagement and Communication	Developing software to enhance public engagement, participation, and communication within smart cities.	Medium	Creating a mobile app for residents to report urban issues, provide feedback, and access city services.
Cybersecurity and Data Privacy	Implementing robust cybersecurity measures and data protection protocols to safeguard sensitive urban data.	High	Developing a comprehensive cybersecurity system to protect smart city infrastructure from cyber threats.
Continuous Improvement	Enabling the continuous development and enhancement of software solutions to adapt to evolving urban needs.	Medium	Implementing an agile development process for a smart city's digital infrastructure, allowing for ongoing improvements.

TABLE 13.3

Key Factors of GSD and Software Examples

Factor Name	Benefits	Impact	Software Example
IoT Platforms	- Real-time monitoring of urban infrastructure. - Efficient resource management. - Enhanced data-driven decision-making.	High	IBM Watson IoT, Microsoft Azure IoT Hub, ThingWorx.
Data Analytics Platforms	- Informed urban planning. - Predictive maintenance. - Improved citizen services.	High	SAS Analytics, Tableau, Apache Hadoop, IBM Cognos.
Smart Grid Management Platforms	- Energy efficiency. - Integration of renewable energy sources. - Load balancing.	High	Siemens Spectrum Power, Schneider Electric EcoStruxure Grid.
Traffic Management and Mobility Platforms	- Reduced congestion. - Improved public transit. - Optimized transportation.	High	Siemens Mobility Management, HERE Mobility, Moovit.
Public Safety and Emergency Management Platforms	- Enhanced emergency response. - Real-time incident monitoring. - Improved disaster management.	High	Motorola PremierOne, Hexagon Safety & Infrastructure.
Waste Management Platforms	- Efficient waste collection. - Cost reduction. - Improved environmental impact.	Medium	RubiconSmart, SmartBin, Compology.
Urban Planning and Design Platforms	- Simulated urban development models. - Sustainable city planning. - Improved land use.	High	Esri ArcGIS Urban, UrbanFootprint, CityEngine.
Digital Services Platforms	- Increased citizen engagement. - Streamlined city service delivery. - Enhanced accessibility.	Medium	Accela Civic Platform, CivicPlus, Granicus.
Open Data Platforms	- Data transparency. - Encouragement of innovation. - Civic participation.	Medium	CKAN, Socrata, OpenDataSoft.
Smart Parking Platforms	- Reduced traffic congestion. - Efficient parking management. - Enhanced user experience.	Medium	Parkmobile, ParkiFi, Streetline.
Building Management Systems (BMS)	- Energy efficiency. - Improved occupant comfort. - Cost savings.	High	Johnson Controls Metasys, Honeywell Building Solutions.
Environmental Monitoring Platforms	- Improved air quality. - Noise pollution control. - Data for sustainability initiatives.	Medium	Acoem (formerly Ecotech), Aeroqual, Libelium.
Healthcare Management Platforms	- Access to healthcare services. - Health data management. - Telehealth solutions.	Medium	Epic Systems, Cerner, eClinicalWorks.
Education and E-Learning Platforms	- Access to educational resources. - Online learning opportunities. - Collaboration tools for educators and students.	Medium	Moodle, Blackboard, Google Classroom.
Tourism and Culture Platforms	- Promotion of cultural events and tourism services. - Enhanced visitor experience. - Local business support.	Medium	Eventbrite, TripAdvisor, Eventzilla.

13.8 CONCLUSION

This chapter discusses how Global Software Development (GSD) can enhance smart cities by promoting innovation, efficiency, sustainability, and inclusivity. As urbanization grows, GSD}s role in shaping future cities will likely expand. However, implementing GSD in smart cities presents unique challenges due to the intricate nature of integrating software solutions with urban infrastructure. Several potential benefits of GSD make it an increasingly common practice in the software industry. However, it requires careful planning and consistent execution. Identifying best practices is crucial – research gaps such as security, management, distance, interaction, and communication impact GSD performance. The literature survey identified several research gaps that impact the performance of GSD, such as security, management, distance, interaction and communication types.

Geographical distance can reduce the likelihood of face-to-face communication, leading to slower decision-making, difficulties in complex discussions, and challenges in building personal relationships. Temporal distance, on the other hand, refers to the difference in time experienced by development teams in GSD, creating problems in coordinating team activities, communication, and collaboration. Sociocultural distance is the variation in values, beliefs and behaviours between groups, which can create gaps in communication. Smart cities need multiple software solutions to function optimally. The chapter presents the software platforms, impact, and usage depending on each smart city project's goals and priorities. The choice of platforms aligns with the city's vision for urban development and the challenges it aims to address.

REFERENCES

Akbar, Muhammad Azeem, and Víctor Leiva. (2024). "A new taxonomy of global software development best practices using prioritization based on a fuzzy system." *Journal of Software: Evolution and Process 36*(3), e2629.

Almalki, Faris A., Saeed H. Alsamhi, RadhyaSahal, Jahan Hassan, Ammar Hawbani, N. S. Rajput, Abdu Saif, Jeff Morgan, and John Breslin.(2023). "Green IoT for eco-friendly and sustainable smart cities: Future directions and opportunities." *Mobile Networks and Applications* 28(1): 178–202.

Alshamaila, Yazn, Savvas Papagiannidis, Hamad Alsawalqah, and Ibrahim Aljarah. (2023). "Effective use of smart cities in crisis cases: A systematic review of the literature." *International Journal of Disaster Risk Reduction* 85: 103521.

Dalal, Sandeep, and Susheela Hooda. (2017). "A novel technique for testing an aspect oriented software system using genetic and fuzzy clustering algorithm." In *2017 International Conference on Computer and Applications (ICCA)*, pp. 90–96. IEEE.

Hooda, Susheela, Vikas Lamba, Sharanpreet Kaur, Vidhu Kiran Sharma, Raju Kumar, and Vandana Sood. (2022). "Impact of software quality on" GA-FC" software testing technique." In *2022 International Conference on Computational Intelligence and Sustainable Engineering Solutions (CISES)*, pp. 142–147. IEEE.

Humayun, Mamoona, Mahmood Niazi, Mohammed Assiri, and Mariem Haoues. (2023). "Secure global software development: A practitioners' perspective." *Applied Sciences.* 13(4): 2465.

Ilyas, Muhammad, Siffat Ullah Khan, Habib Ullah Khan, and Nasir Rashid. (2023). "Software integration model: An assessment tool for global software development vendors." *Journal of Software: Evolution and Process* 36(4): e2540.

Khan, Abdul Wahid, Shah Zaib, Faheem Khan, Ilhan Tarimer, Jung TaekSeo, and Jiho Shin. (2022). "Analyzing and evaluating critical cyber security challenges faced by vendor organizations in software development: SLR based approach." *IEEE Access* 10: 65044–65054.

Mishra, Debasisha. (2023). "Developing a knowledge-based perspective of coordination in global software development." *VINE Journal of Information and Knowledge Management Systems.* https://doi.org/10.1108/VJIKMS-08-2022-0270

Noor, Maham, and Zeeshan Ali Rana. (2018). "Towards better knowledge management in global software engineering." In *2018 4th International Conference on Computer and Information Sciences (ICCOINS)*, pp. 1–6. IEEE.

Sajjad, Muhammad, Mansoor Nasir, Khan Muhammad, Siraj Khan, Zahoor Jan, Arun Kumar Sangaiah, Mohamed Elhoseny, and Sung Wook Baik. (2020). "Raspberry Pi assisted face recognition framework for enhanced law-enforcement services in smart cities." *Future Generation Computer Systems* 108: 995–1007.

Siddique, Saba, Muhammad Naveed, Atif Ali, Ismail Keshta, Muhammad Islam Satti, Azeem Irshad, Zakaria Alomari, Onome Christopher Edo, and Oladapo Ayodeji Diekola. (2023). "An effective framework to improve the managerial activities in global software development." *Nonlinear Engineering* 12(1): 20220312.

Statistics. "Worldwide-smart-city-market-revenue", 1111626 (2023). https://www.statista.com/statistics/1111626/worldwide-smart-city-market-revenue/

Su, Yiyi, and Di Fan. (2023). "Smart cities and sustainable development." *Regional Studies.* 57(4): 722–738.

Yan, Zheming, Zao Sun, Rui Shi, and Minjuan Zhao. (2023). "Smart city and green development: Empirical evidence from the perspective of green technological innovation." *Technological Forecasting and Social Change* 191: 122507.

Yousif, Jabar H., and Dinesh Kumar Saini. (2020). "Big data analysis on smart tools and techniques." *Cyber Defense Mechanisms*, 111–130. 10.1201/9780367816438-8

14 Applications
Smart City and Urbanization Systems

Manoj Kumar Mahto
Vignan Institute of Technology and Science, Hyderabad, India

Durgesh K. Srivastava
Chitkara University, Punjab, India

Mahender Singh Kaswan
Lovely Professional University, India

14.1 INTRODUCTION

Amidst an unprecedented global wave of urbanization, where over half of the world's population now resides in cities (United Nations, 2018), urban areas grapple with a multitude of complex challenges. From the daily throes of traffic congestion and environmental pollution to the intricacies of energy consumption, resource management, and public health, cities stand at the intersection of numerous pressing issues. In response, the concept of smart cities has emerged as a visionary solution. It envisions a future where advanced technologies and innovative strategies collaborate to craft urban environments that are not just sustainable and efficient but fundamentally liveable (Caragliu et al., 2011). Embarking on a comprehensive exploration, we delve into the realm of smart cities and urbanization systems. Characterized by their embrace of digital technologies, utilization of data analytics, and integration of interconnected infrastructure, these systems aim to uplift the quality of life for residents. The overarching goal is clear: to optimize the allocation of resources, minimize environmental impacts and ultimately create cities where living is an art form. This journey encompasses diverse areas, including smart transportation, which leverages IoT sensors, autonomous vehicles, and intelligent traffic management to alleviate congestion and reduce emissions (Zheng et al., 2019). Energy management is also a focal point, with the integration of renewable sources and smart grids reshaping energy consumption (Capozzoli et al., 2021). Smart waste management, healthcare innovations like telemedicine, data-driven public safety measures and citizen engagement platforms contribute to safer, more connected urban environments. However, privacy concerns, data security, the digital divide and substantial financial investments present challenges to the realization of smart city potentials (Albino

DOI: 10.1201/9781003467892-14

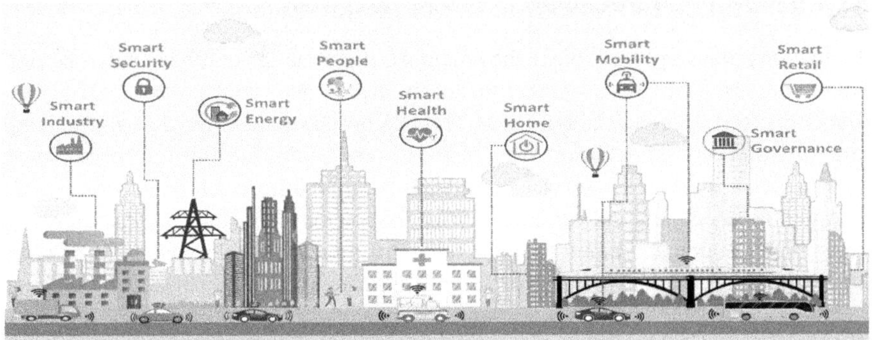

FIGURE 14.1 Exploring the transformative application of smart cities in the dynamic landscape of urbanization (Inxee, 2023).

et al., 2015). This chapter serves as a guide to exploring the current applications and future prospects of these innovations, offering insights into their transformative impact on cities worldwide and their role in shaping a more connected, prosperous and sustainable urban future. Figure 14.1: Applications of smart city technologies in enhancing urban infrastructure and services.

14.1.1 CONCEPT AND GOALS OF SMART CITIES

Imagine a city where innovation isn't just a buzzword but a way of life. That's the essence of smart cities – a transformative approach to urban development that goes beyond bricks and mortar. At its core, the goal is simple yet profound: to make life better for the people who call these cities home. It's about creating spaces that not only thrive economically but also nurture the well-being of residents and care for the environment. In the tapestry of urban life, smart cities weave a story of progress, where every innovation is a step towards a brighter, more sustainable future. Smart cities prioritize the optimization of urban services, such as transportation, healthcare and education, to make them more efficient and citizen-centric, ultimately creating environments that offer a higher quality of life. Additionally, these cities recognize the role of innovation and entrepreneurship in driving economic prosperity, fostering an ecosystem that attracts investments and promotes job creation. Environmental sustainability is a fundamental goal, achieved through the integration of renewable energy sources, smart grids, and energy-efficient infrastructure, as well as sustainable urban planning and waste management practices. Efficient resource allocation, facilitated by data analytics and real-time monitoring, contributes to cost savings and resource conservation, exemplified by optimized transportation systems, intelligent energy grids, and streamlined waste management (Nam and Tan, 2016). Furthermore, smart cities prioritize inclusivity and equity, bridging the digital divide, promoting social cohesion, and ensuring that all segments of the population have a voice in urban planning (Caragliu et al., 2011). In essence, smart cities signify a paradigm shift in urban development, underpinned by technology, data and a comprehensive commitment to improving residents' well-being while simultaneously fostering economic growth and environmental sustainability.

14.1.2 Role of Technology in Urban Transformation

Technology plays a pivotal role in the ongoing transformation of urban landscapes into smart and sustainable cities, underpinned by digital innovations, data analytics and interconnected systems (Caragliu et al., 2011). Through data-driven decision-making, cities harness vast datasets from diverse sources, enabling real-time informed choices for resource allocation and service optimization. Smart infrastructure, including grids, transportation systems and buildings, forms the foundation of these cities, improving energy distribution and mobility while reducing environmental impact (Zheng et al., 2019). High-speed connectivity, driven by broadband and 5G networks, facilitates seamless communication, underpinning various smart city initiatives (Zhang et al., 2018). Technology also enhances governance through e-government platforms, promoting transparency and citizen engagement. These innovations contribute to sustainability goals through energy-efficient buildings, smart waste management and green infrastructure. Moreover, technology strives for inclusivity by bridging the digital divide and ensuring equitable access (Nam & Tan, 2016). However, challenges such as data privacy, cybersecurity and digital exclusion warrant ongoing attention to fully harness technology's potential (Albino et al., 2015).

14.2 SMART TRANSPORTATION

Envision a city where smart transportation transforms the way we move, a pivotal element in the broader smart city vision. It harnesses advanced technologies and data-driven strategies to redefine urban mobility (Zheng et al., 2019). This comprehensive approach involves intelligent traffic management systems utilizing real-time data to optimize traffic flow, autonomous vehicles relying on sensors and AI for safe navigation, and mobility-as-a-service platforms streamlining various transportation modes (Li et al., 2017). Additionally, smart parking solutions equipped with sensors and mobile apps ease parking spot location, electric vehicles and charging infrastructure reduce emissions, and predictive maintenance enhances the reliability of public transportation (Xie et al., 2017; Pulido et al., 2019). Challenges including data privacy, cybersecurity and regulatory considerations for autonomous vehicles must be addressed, alongside ensuring equitable access for all segments of the population (Chien & Ding, 2019). In essence, smart transportation aims to reduce congestion, enhance safety, and promote efficiency and sustainability in urban transportation systems. Figure 14.2: Overview of smart transportation systems in smart cities, highlighting the integration of technology for improved efficiency and connectivity.

14.2.1 Intelligent Traffic Management Systems

Smart transportation hinges on Intelligent Traffic Management Systems (ITMS), employing real-time data and cutting-edge technologies to streamline traffic flow, ease congestion, and elevate road safety standards (Li et al., 2017). These systems deploy an array of sensors, cameras, and data analytics to monitor live traffic conditions. This enables transportation authorities to swiftly grasp the current state of the road network, thanks to immediate insights derived from these real-time observations. ITMS includes adaptive traffic signal control, dynamically adjusting

FIGURE 14.2   The evolution of smart transportation systems in the context of smart cities (Steve, 2020).

signal timing to reduce wait times at intersections and minimize idling, ultimately decreasing travel times for commuters. Furthermore, ITMS can detect and predict congestion, facilitating proactive interventions such as rerouting traffic and providing real-time updates to drivers, thus improving overall road efficiency (Ding et al., 2019). This approach not only reduces environmental impact by minimizing fuel consumption and emissions but also enhances road safety through accident detection and emergency response notifications (Li et al., 2017). Nevertheless, challenges such as infrastructure costs, data privacy concerns, and the necessity for multi-stakeholder collaboration must be addressed, along with a focus on ensuring equitable access to ITMS benefits for all citizens. In essence, ITMS contributes significantly to more efficient urban mobility, reduced environmental impact and improved overall quality of life for city residents within the framework of smart cities.

14.2.2   Connected and Autonomous Vehicles

Connected and Autonomous Vehicles (CAVs) represent a technological revolution poised to transform urban transportation within smart cities, offering a spectrum of advantages and intricate challenges. Outfitted with sophisticated sensors, artificial intelligence and seamless real-time connectivity, these vehicles hold the promise of markedly improving road safety through a reduction in human errors (Litman, 2018). Beyond that, they possess the capability to transform urban traffic management by creating collaborative networks that fine-tune traffic flow, diminishing congestion, and elevating the overall efficiency of transportation. CAVs also promise increased accessibility, providing those with disabilities and the elderly newfound mobility (Keates, 2019). In terms of environmental sustainability, they can contribute to reduced fuel consumption and emissions through efficient driving patterns, and the transition to electric autonomous vehicles further aligns with smart city environmental goals (Litman, 2018). However, their integration into smart cities comes with cybersecurity concerns, given their extensive connectivity, and challenges surrounding liability allocation in accidents, necessitating careful consideration (Kambatla et al., 2019). In essence, CAVs are poised to revolutionize urban mobility, yet their successful integration hinges on addressing these multifaceted challenges.

14.2.3 Multimodal Transportation Solutions

Multimodal transportation solutions are a cornerstone of smart city mobility, offering integrated and versatile transportation modes that enable seamless travel transitions for urban residents (Bieńkowski et al., 2018). These systems dismantle the divides among different transportation modes, encompassing buses, trains, bicycles, ride-sharing and walking. They are frequently facilitated through digital platforms and mobile apps that furnish real-time information. Multimodal solutions seek to encourage the use of public transit, cycling, and walking with the goal of curbing road congestion, enhancing air quality, and reducing energy consumption in urban areas. Sustainability is a key aspect, encouraging eco-friendly transportation choices (Bieńkowski et al., 2018). Furthermore, these systems enhance accessibility for a wide range of individuals, including those with disabilities, the elderly, and low-income residents who rely on diverse transportation options. Multimodal transportation relies on data-driven decision-making to optimize services and infrastructure (Sivakumar et al., 2018). However, their successful implementation hinges on effective urban planning, investment, and equitable access to transportation choices for all citizens (Litman, 2020). In essence, multimodal transportation solutions hold the potential to transform urban mobility, promoting sustainability and accessibility within smart cities.

14.2.4 Reducing Congestion and Improving Mobility

Efficiently reducing traffic congestion and enhancing overall mobility stand as pivotal objectives within the realm of smart city transportation systems (Litman, 2019). Beyond mere productivity losses, congestion contributes to environmental challenges, including heightened greenhouse gas emissions and the deterioration of air quality. Smart cities are employing an array of strategies and cutting-edge technologies to tackle these pressing urban issues. These include the implementation of advanced traffic management systems capable of optimizing signal timings and offering real-time adaptive routing suggestions. Simultaneously, they prioritize the expansion and enhancement of public transit networks, encouraging residents to embrace efficient and reliable alternatives to private vehicles. Ridesharing and carpooling platforms have become integral components of urban mobility, effectively reducing the number of single-occupancy vehicles on the road. To promote sustainable and healthy transportation, smart cities invest in infrastructure improvements like dedicated bike lanes and pedestrian-friendly pathways, encouraging walking and cycling. The integration of Intelligent Transportation Systems (ITS) facilitates real-time data exchange, enabling efficient traffic management and adaptive signal control. Moreover, demand management strategies, notably congestion pricing and flexible work hours, aim to evenly distribute traffic throughout the day, mitigating peak-hour congestion. However, the successful implementation of these strategies necessitates overcoming significant challenges, including securing substantial investments in transportation infrastructure, gaining public acceptance of new mobility solutions, and ensuring equitable access to the benefits of congestion reduction (Litman, 2019). At its core, smart cities are at the forefront of pioneering inventive approaches to bring about urban mobility that is not just efficient and sustainable but also easily accessible. This commitment stems from a dedication to improving the quality of life for residents, all while tackling crucial environmental and congestion challenges.

14.3 ENERGY MANAGEMENT

Energy management is a cornerstone of smart city development, strategically optimizing energy production, distribution, and consumption while fostering sustainability and resilience within urban landscapes (Balderrama-Subieta et al., 2017). In the heart of smart cities lies a commitment to weaving sustainability into their energy tapestry. They place a premium on incorporating sources like solar, wind, and hydropower, not just to cut down on greenhouse gas emissions but also to fortify energy security. At the core of this effort are smart grids, orchestrating a ballet of real-time monitoring and optimization in electricity distribution. This not only boosts grid reliability but also seamlessly integrates renewable energy into the mix (Li et al., 2013). Energy efficiency is paramount, with smart cities investing in energy-efficient infrastructure, including automated systems in smart buildings, leading to cost savings and reduced environmental impact (Capozzoli et al., 2021). Demand response initiatives encourage consumers to adapt their energy consumption patterns, reducing peak demand and enhancing grid stability. Energy storage solutions, like cutting-edge batteries, play a crucial role in ensuring a steady and dependable energy supply, especially when it comes to the unpredictable nature of renewable sources. To make things even more fascinating, we turn to the power of data analytics and IoT technologies. They work hand in hand to keep a watchful eye on how we consume energy, providing the insights needed for wise and informed decision-making. It's not just about keeping the lights on; it's about doing so in a way that's smart, efficient, and tailored to our energy needs. Challenges encompass high upfront costs of renewable infrastructure, the need for effective energy storage, regulatory complexities, and the promotion of energy-efficient behaviours among residents and businesses (Jin et al., 2016). In essence, comprehensive energy management strategies are fundamental to smart cities' sustainability efforts, striving to mitigate environmental impact, enhance energy security and promote resilience in urban environments.

14.3.1 SMART GRID TECHNOLOGIES

Smart grid technologies constitute a pivotal element of energy management in smart cities, representing a transformative shift in how electricity is generated, distributed and consumed. These grids involve the modernization of traditional electrical systems through the incorporation of digital technologies, including advanced sensors, meters and communication networks, enabling real-time data collection and analysis. Central to their functionality is the deployment of Advanced Metering Infrastructure (AMI), or smart meters, which furnish detailed insights into electricity consumption, empowering both consumers and utilities to monitor and optimize energy usage efficiently. Think of smart grids as the guardians of a sustainable energy ballet. They don't just welcome Distributed Energy Resources (DERs) like solar panels and wind turbines; they invite them to a localized energy party, even letting us share any extra power we have back to the grid. These grids are like wizards, magically spotting and fixing issues on their own, ensuring we hardly notice a blip during outages and swiftly bringing the lights back on. But here's the twist: they're not just about technology; they're about us. Smart grids encourage a dance of sorts, urging us to tweak our energy habits when the grid needs a breather, offering perks for doing so and ultimately lessening the need for extra power generation (Hao et al., 2019). It's like having a conversation with our energy system, where the grid listens and responds to keep the show going smoothly. Furthermore, smart grids promote energy efficiency by identifying opportunities to curtail energy wastage and optimize distribution, resulting in cost savings for utilities and reduced energy bills for consumers. Challenges in their implementation include addressing data security and privacy concerns, securing substantial infrastructure investments, navigating regulatory complexities, and ensuring equitable access to the benefits they offer (Lu et al., 2017). In essence, smart grid technologies epitomize the modernization and resilience of energy management systems in smart cities, underpinning sustainability and efficiency while demanding solutions to multifaceted challenges. Figure 14.3: Illustration of smart grid technologies for energy management in smart cities, showcasing systems that enhance energy efficiency and reliability.

14.3.2 Renewable Energy Integration

Renewable energy integration stands as a foundational pillar of smart city energy management, representing a proactive response to pressing environmental and sustainability challenges. Smart cities are increasingly harnessing a diverse array of renewable energy sources, including solar power through photovoltaic panels, wind energy from urban wind farms, hydropower from local water flows, and bioenergy derived from organic waste (Pandey et al., 2016). These initiatives not only reduce greenhouse gas emissions but also contribute to a diversified and cleaner energy mix, effectively mitigating the reliance on fossil fuels. Imagine a symphony where energy storage solutions are the dependable backup singers, ensuring the show goes on seamlessly. These solutions, like advanced batteries and pumped hydro storage, step up when renewable energy takes a breather, bridging those intermittent gaps in generation. And then, there are smart grids – imagine them as the conductors of this symphony, fuelled by digital technologies. They weave a dynamic framework, effortlessly integrating renewable sources into the electrical grid. It's not just about delivering energy; it's a real-time dance of monitoring, control and optimization, ensuring that the lights stay on and the energy flow is as harmonious as a well-coordinated orchestra. Nevertheless, addressing challenges related to intermittency, energy storage, grid management and policy support is essential for the successful

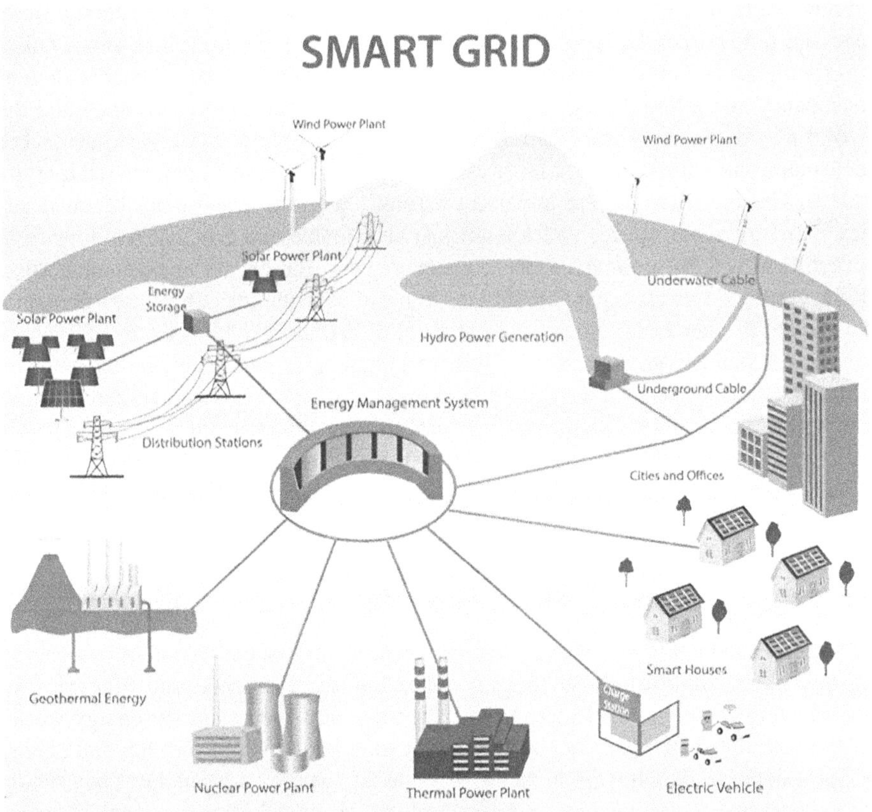

FIGURE 14.3 Harnessing smart grid technologies for sustainable energy management within the framework of smart cities (Joy, 2022).

and sustainable adoption of renewable energy solutions within smart cities (Dutta et al., 2019). This multifaceted approach not only enhances the environmental footprint of urban areas but also contributes to their energy resilience and sustainability, marking a critical step near an olive green and more sustainable future for smart cities and their inhabitants.

14.3.3 Demand Response and Energy Efficiency

Demand response and energy efficiency strategies are pivotal components of smart city energy management, poised to revolutionize the way urban areas consume and manage energy resources. Demand response initiatives empower consumers to actively engage with their energy consumption patterns, enabling them to respond to grid conditions and price signals by voluntarily reducing or shifting electricity use during peak periods. This not only contributes to grid stability and reduced energy costs but also promotes a more sustainable energy ecosystem. Energy efficiency measures encompass an array of technologies and practices, from energy-efficient appliances to sophisticated building automation systems, that minimize energy wastage and optimize consumption (Capozzoli et al., 2021). By equipping smart buildings with automated systems, energy usage is finely tuned based on factors such as occupancy and external conditions, resulting in substantial energy savings. Enter the world of smart meters, the wizards that offer a real-time glimpse into our energy consumption, adding an extra layer of effectiveness to these initiatives (Hao et al., 2019). These endeavours are in perfect harmony with the grand goals of smart cities – to craft urban spaces that are not just sustainable and efficient but downright liveable. How? By pulling us, the consumers, into the energy management spotlight, cutting down costs, and making meaningful contributions to the environmental sustainability dance. It's not just about data; it's about involving us in the journey towards a smarter, greener future. However, challenges persist, including the need for advanced infrastructure, regulatory support, and fostering behavioural changes among residents and businesses to fully embrace energy-efficient practices. Even in the face of challenges, the incorporation of demand response and energy efficiency stands at the core of the smart city vision, steering the course towards a greener and more sustainable urban future.

14.3.4 Microgrids and Energy Storage

Microgrids and energy storage systems are central components of smart city energy management, significantly enhancing the reliability, resilience and efficiency of urban energy ecosystems. Enter microgrids, those superheroes of the energy world. Picture localized energy systems armed with solar panels, wind turbines and energy storage devices, standing tall in the face of grid outages or emergencies. They swoop in to guarantee an uninterrupted power supply to essential facilities, all while lending a hand to grid stability and the seamless integration of renewable energy. And that's not all; energy storage systems, featuring the likes of batteries and pumped hydro storage, step into the spotlight. They play a crucial role in taming the unpredictability of renewables, juggling energy usage to perfection, and curbing peak demand. The result? Lower energy costs and a thriving ecosystem of sustainability (Hao et al., 2019). The integration of these systems into smart grids, facilitated by advanced monitoring and control technologies, is essential for seamless coordination within the broader electrical grid. While their deployment aligns with the core objectives of smart cities, including enhanced reliability and sustainability, challenges such as initial costs, technical complexity, and regulatory considerations must be addressed to fully realize their potential.

14.4 WASTE MANAGEMENT

Advanced waste management practices are indispensable for smart cities aiming to enhance sustainability and environmental responsibility. These cities employ smart waste collection methods, utilizing sensor-equipped bins to monitor waste levels, optimize collection routes, and minimize costs (Korhonen et al., 2018). Additionally, waste sorting and recycling are facilitated through both citizen engagement and automated facilities that employ robotics and artificial intelligence for efficient material

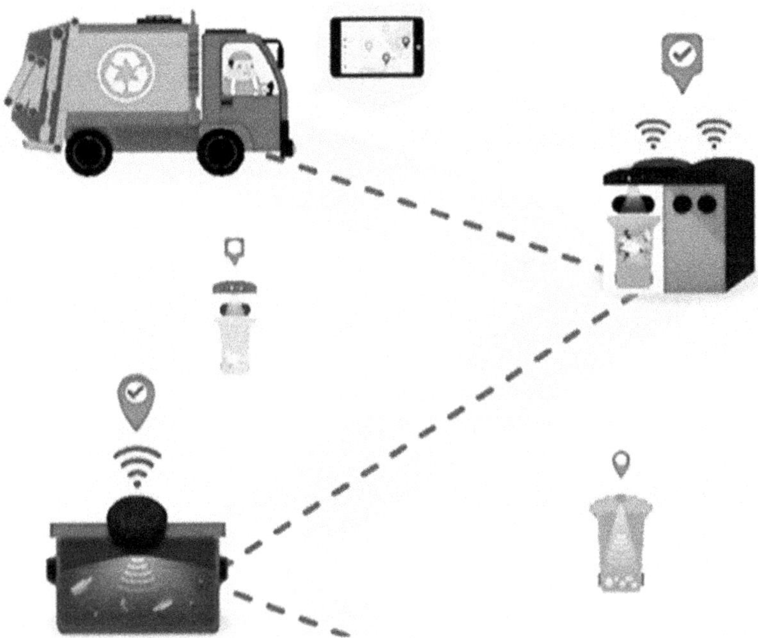

FIGURE 14.4 Streamlining urban sustainability with smart waste management systems in smart cities (IoT Misr, 2020).

separation (Zhang et al., 2018). Waste-to-energy conversion technologies play a pivotal role in transforming organic waste into renewable energy, thus reducing landfill use and promoting resource efficiency (Arena et al., 2019). Real-time monitoring of waste processes empowers data-driven decision-making and proactive responses to waste management challenges, aligning with smart city goals of sustainability and liveability. However, challenges related to initial investments, technological infrastructure and public awareness must be addressed to fully realize the benefits of advanced waste management in smart cities. Figure 14.4: Depiction of smart waste management systems in smart cities, highlighting innovative solutions for efficient waste collection and processing.

14.4.1 Smart Waste Collection and Recycling

Efficient waste collection and recycling stand as crucial pillars of sustainability within smart cities. Picture this urban landscape where technology takes centre stage in a rather clever act. Enter smart bins, decked out with sensors that keep a real-time tab on waste levels. It's like having bins that can talk! And what do they say? They whisper insights to the waste collection wizards, helping them map out the most efficient routes and schedules. This isn't just about saving time and money; it's a dance of efficiency that lessens the environmental footprint of those waste collection vehicles. It's technology doing a little jig to make our cities cleaner and greener. Citizen engagement is actively promoted through digital platforms, enabling residents to access information on waste sorting, recycling locations and environmentally responsible disposal options, thereby fostering a heightened sense of environmental consciousness. Moreover, recycling facilities incorporate advanced sorting technologies like robotics and artificial intelligence, enhancing recycling rates and minimizing contamination of recyclables (Arena et al., 2019). In line with circular economy principles, smart cities embrace waste reduction and recycling as a means to minimize waste generation, lower landfill usage, and promote resource conservation (Korhonen et al., 2018). Embracing these holistic approaches in waste management isn't just about taking out the trash; it's a thoughtful strategy that dances with efficiency, cuts down on carbon emissions, and champions the idea of a circular economy. It's like giving waste a second chance at being something useful. In the grand scheme of things, these practices align seamlessly with the bigger goals of smart cities – crafting urban havens that are not just sustainable but also radiate an eco-friendly charm. It's a nod to a future where our cities are not just smart but also deeply connected to the well-being of our planet.

14.4.2   IoT-enabled Waste Bins

In the realm of smart waste management systems in cities, IoT-enabled waste bins emerge as a transformative technology. These silos are prepared with sensors that continuously monitor waste levels and transmit real-time data to a central management system, allowing for optimized waste collection routes, reduced fuel consumption, and lower emissions. Predictive maintenance features detect and address issues promptly, ensuring the bins remain in good working order. Think of these bins as more than just receptacles – they're storytellers. The data they churn out isn't just about managing waste efficiently; it's a narrative of how we generate waste, a tale told through bits and bytes. This data-driven insight isn't confined to back-end operations; it's a tool for making decisions that resonate with the pulse of waste generation. But here's the real magic: it's not just about data; it's about us. It sparks conversations with citizens through mobile apps and online platforms, turning waste management into a collaborative dance, where everyone's steps matter. The implementation of IoT-enabled waste bins results in significantly optimized waste collection processes, reduced operational costs, lower environmental impact and enhanced citizen participation, aligning with the sustainability goals of smart cities. However, challenges such as initial investments, data security, and the need for public awareness and involvement must be addressed to fully harness their potential.

14.4.3   Data-Driven Waste Reduction Strategies

Data-driven waste reduction strategies are pivotal within smart cities, helping to mitigate waste generation, enhance recycling, and reduce environmental impacts. Smart cities utilize data analytics to gain insights into waste generation patterns, sources, and composition, facilitating the development of targeted waste reduction initiatives. These initiatives encompass behavioural change campaigns, inspired by data-driven insights, aimed at encouraging residents and businesses to adopt eco-friendly practices like composting and reducing single-use plastics. Moreover, data-driven strategies inform dynamic waste collection schedules, optimizing routes and ensuring bins are emptied only when necessary, effectively reducing unnecessary pickups and associated emissions (Arena et al., 2019). Recycling education efforts are also guided by data analysis, enabling cities to pinpoint areas with lower recycling rates and target campaigns to improve recycling practices. Furthermore, data-driven approaches optimize waste-to-energy conversion processes, maximizing energy recovery potential from waste by analysing waste stream compositions (Korhonen et al., 2018). These strategies yield numerous benefits, including efficient waste management, reduced landfill usage, lower carbon emissions, and resource conservation, all in line with the sustainability objectives of smart cities. Nonetheless, challenges such as the necessity for robust data collection infrastructure, data privacy concerns and the importance of public engagement in waste reduction efforts must be effectively addressed to ensure the success of data-driven waste reduction initiatives.

14.4.4 Circular Economy Approaches

Circular economy principles have gained prominence in smart cities' efforts to achieve sustainable waste management and resource utilization. These approaches prioritize resource recovery and the reuse of materials extracted from waste streams. Smart cities deploy innovative technologies to extract valuable resources like metals, plastics, and organic matter for reuse in various industries. Extended Producer Responsibility (EPR) policies, which shift product management and recycling responsibilities from end-users to producers, are integral to circular economy approaches. Smart cities collaborate with manufacturers to design products that are easily disassembled and recycled. Furthermore, initiatives like Product-as-a-Service (PaaS) models promote product longevity and reduce waste by offering products as services rather than ownership (Tukker et al., 2019). These initiatives encourage residents to access products on demand, minimizing the disposal of short-lived items. Circular economy strategies in smart cities also emphasize waste prevention through the design of durable and repairable products. They promote a culture of repair and reuse, supporting initiatives such as repair cafés and community-based repair programs. These approaches yield benefits such as resource conservation, reduced landfill waste, lower greenhouse gas emissions, and economic opportunities through resource recovery (Geng et al., 2019). Challenges include transitioning from a linear to a circular economy, regulatory frameworks, and the need for consumer behaviour change (Tukker et al., 2019). Smart cities address these challenges through policy development and public awareness campaigns. To sum it up, adopting circular economy approaches is paramount for fostering sustainable waste management and resource utilization in smart cities. This commitment contributes significantly to cultivating a more resilient and environmentally responsible urban environment.

14.5 HEALTHCARE

Healthcare in smart cities is a vital component of urban transformation, driven by the need to provide efficient and accessible healthcare services in the face of urban population growth. Smart cities utilize advanced technologies and data-driven approaches to address these challenges, encompassing telemedicine and remote healthcare services, wearable health technologies for real-time health monitoring, predictive healthcare through data analytics and AI, and data-driven public health management. These innovations promise to enhance healthcare accessibility and quality. However, the implementation of healthcare in smart cities also comes with challenges, including data security, patient privacy, interoperability of healthcare systems and addressing the digital divide. Effective governance and regulations are crucial to overcome these challenges and ensure equitable healthcare access (Kouroubali et al., 2019).

14.5.1 Telemedicine and Remote Patient Monitoring

Telemedicine and remote patient monitoring stand as cornerstone elements within the realm of healthcare in smart cities, reshaping the delivery and accessibility of medical services. Telemedicine, employing digital technologies and telecommunications, facilitates remote consultations between patients and healthcare providers, significantly enhancing healthcare accessibility, particularly in underserved or remote areas. Patients can receive medical consultations, diagnoses, and even treatments from the comfort of their homes, diminishing the necessity for physical visits to healthcare facilities. Complementing telemedicine, remote patient monitoring enables continuous tracking of patients' vital signs and health data through wearable devices and IoT sensors, offering real-time data transmission to healthcare professionals (Vegesna et al., 2017). This proactive approach allows for timely interventions and early detection of health issues, resulting in improved patient care and reduced healthcare system burdens by preventing hospital admissions and readmissions. Notably, these technologies contribute to better patient outcomes and are especially beneficial in managing chronic conditions like diabetes and hypertension (Vegesna et al., 2017). Nevertheless, the effective integration of telemedicine and remote patient monitoring in smart cities requires addressing challenges related to data privacy, security, and the imperative for seamless interoperability among various healthcare systems and devices. Achieving a balance between leveraging the potential of these technologies and safeguarding patient information is critical for their widespread adoption and their effectiveness in enhancing healthcare delivery. Figure 14.5: Illustration of telemedicine and remote patient monitoring in smart cities, emphasizing the role of technology in enhancing healthcare accessibility and efficiency.

14.5.2 Health Data Analytics for Disease Surveillance

In the healthcare landscape of smart cities, the proactive monitoring and management of public health take centre stage, with health data analytics playing a crucial role in disease surveillance. Through the integration of extensive datasets from diverse sources – ranging from electronic health records to wearable health technologies and real-time environmental data – advanced data analytics and artificial intelligence (AI) models step into the spotlight. Their mission: to identify disease outbreaks, trace the spread of infections, and evaluate trends in population health. These analytical insights empower healthcare authorities to make informed decisions

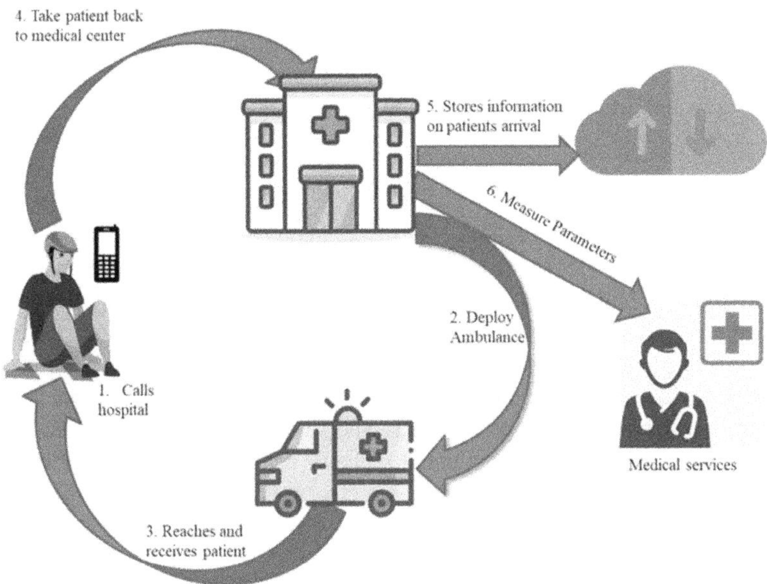

FIGURE 14.5 Advancing healthcare access through telemedicine and remote patient monitoring in smart cities (Poongodi et al., 2021).

swiftly, allocate resources effectively, and implement targeted public health interventions. Additionally, predictive analytics models can forecast disease trends, allowing for timely responses and the optimization of healthcare resources. However, the successful application of health data analytics for disease surveillance also entails addressing challenges such as data privacy, security and the need for data interoperability among diverse healthcare systems (Chen et al., 2018). Striking a balance between the utilization of these powerful analytics tools and safeguarding patient privacy is paramount for their continued success in enhancing public health surveillance within smart cities.

14.5.3 SMART HOSPITALS AND HEALTHCARE INFRASTRUCTURE

Smart hospitals and healthcare infrastructure constitute a cornerstone of healthcare in smart cities, introducing transformative changes in patient care, operational efficiency, and resource management. These advanced healthcare facilities leverage cutting-edge skills, such as the Internet of Things (IoT), real-time monitoring, and AI, to enhance patient experiences and outcomes. IoT devices embedded in smart hospitals collect real-time patient data, enabling remote monitoring of vital signs, medication adherence, and early detection of healthiness issues. AI-driven predictive analytics assist in optimizing hospital workflows, patient scheduling, and resource allocation, ultimately reducing operational costs and enhancing healthcare delivery (Tuli et al., 2020). Furthermore, the integration of electronic health records (EHRs) and telemedicine capabilities ensures seamless patient information exchange and remote consultations, improving healthcare accessibility and reducing the need for physical visits. Nonetheless, the adoption of smart hospitals faces challenges, including data security, interoperability, and the need for healthcare professionals' digital literacy. Addressing these concerns is vital to fully realize the potential benefits of smart healthcare infrastructure in urban settings.

14.5.4 IMPROVING PUBLIC HEALTH OUTCOMES

Enhancing public health outcomes is a fundamental objective within the framework of healthcare in smart cities, driven by innovative strategies and technologies. Data analytics and AI play pivotal roles in monitoring and improving population health. Imagine a city that not only cares for its residents but also understands their health like a trusted friend. By pulling together information from electronic health records, wearable devices and environmental sensors, this city becomes a vigilant guardian. It doesn't just see data; it interprets trends, spots health disparities, and, like a caring ally, puts in place interventions tailored to specific needs (Mandl et al., 2020). It's not just about information; it's about creating a city that actively looks out for the well-being of its people. Predictive modelling enables the forecasting of disease outbreaks and healthcare resource needs, ensuring timely responses to health crises. Moreover, interactive citizen engagement platforms and health apps promote healthy lifestyles, encourage preventive care, and facilitate early disease detection, empowering individuals to actively participate in their health management. However, challenges related to data privacy, equity in healthcare access and the digital divide must be addressed to ensure that the benefits of smart public health initiatives are accessible to all residents.

14.6 PUBLIC SAFETY

Public safety stands as a paramount concern in the context of smart cities, and it is at the forefront of urban transformation efforts. Smart cities, through the integration of cutting-edge technologies, data analytics, and interconnected systems, strive to elevate the safety and security of their residents. Real-time data from sensors and surveillance cameras enable data-driven law enforcement, predictive policing, and real-time crime mapping, allowing authorities to respond swiftly to incidents and allocate resources effectively (Bogar, 2017). Interactive citizen engagement platforms facilitate communication between residents and law enforcement agencies, fostering safer and more connected urban environments. Nevertheless, the implementation of public safety initiatives in smart cities entails addressing challenges related to data privacy, ethical concerns, and ensuring equitable access to security measures, emphasizing the need for responsible governance and community involvement.

14.6.1 SURVEILLANCE AND MONITORING SYSTEMS

Surveillance and monitoring systems represent a cornerstone of public safety initiatives within smart cities, harnessing advanced technologies and data-driven approaches to enhance security and response capabilities. These systems encompass a wide array of sensors, cameras, and data analytics tools that enable real-time monitoring of urban areas. Data from these sources are leveraged for predictive policing, crime detection

FIGURE 14.6 Vigilance and monitoring systems enhancing security in smart cities (Wai, 2022).

and incident response, allowing law enforcement agencies to allocate resources effectively and respond swiftly to emergencies. Moreover, these systems contribute to the creation of safer urban environments by acting as deterrents to criminal activities and ensuring a quick response to security threats (Al-Hader et al., 2016). However, the proliferation of surveillance and monitoring systems also raises concerns about individual privacy, data security and the potential for misuse, necessitating careful governance and oversight to strike a balance between security and civil liberties. Figure 14.6: Overview of surveillance and monitoring systems in smart cities, highlighting the use of advanced technologies to enhance public safety and security.

14.6.2 Emergency Response and Disaster Management

Emergency response and disaster management represent critical facets of public safety in smart cities, where technological advancements and data-driven approaches are employed to mitigate risks and improve response effectiveness. Harnessing real-time data from diverse sources such as sensors, social media, and geographic information systems (GIS), smart cities can promptly evaluate the impact of disasters and facilitate swift coordination. Predictive analytics and modelling aid in disaster preparedness and resource allocation, enabling authorities to anticipate scenarios, plan evacuation routes, and deploy resources efficiently (Gupta and Singla, 2016). Furthermore, mobile apps and communication platforms facilitate real-time information dissemination to residents during emergencies, enhancing community engagement and reducing response times. Nevertheless, the success of emergency response and disaster management in smart cities hinges on addressing challenges such as data interoperability, ensuring the resilience of critical infrastructure, and fostering community resilience through public awareness and education.

14.6.3 Predictive Policing and Crime Prevention

Predictive policing and crime prevention are integral components of public safety initiatives in smart cities, harnessing data analytics and machine learning to enhance law enforcement efforts. By analysing historical crime data, socioeconomic factors and environmental variables, predictive policing models can identify high-risk areas and times for criminal activities (Mohler et al., 2015). Law enforcement agencies can proactively allocate resources to deter criminal behaviour and respond swiftly to incidents, ultimately reducing crime rates. Additionally, these technologies facilitate community policing by promoting transparency and engaging residents in crime prevention efforts through interactive platforms. Yet, to ensure the responsible and equitable implementation of predictive policing strategies, ethical concerns surrounding data privacy, fairness and transparency must be duly addressed.

14.6.4 Enhancing Overall Urban Security

A core goal of public safety initiatives in smart cities is to bolster overall urban security. This mission is fuelled by the seamless integration of advanced technologies and data-driven strategies. Smart cities employ a multitude of measures, including video surveillance, sensor networks, biometric identification systems, and data analytics, to monitor and secure critical infrastructure, public spaces, and transportation networks (Orozco-Arroyave et al., 2018). These technologies enable real-time threat detection, rapid response to security incidents and the implementation of robust access control measures, fostering safer urban environments. However, the effective deployment of such security measures necessitates vigilant contemplation of ethical and confidentiality concerns, ensuring that the benefits of enhanced urban security do not infringe upon individual rights.

14.7 CITIZEN ENGAGEMENT

Citizen engagement is a pivotal aspect of smart cities, fostering active contribution and association between residents and urban authorities to co-create more inclusive and responsive urban environments. Digital platforms, mobile apps and interactive tools enable residents to afford response, report problems, and enter data on urban services, promoting transparency and accountability (Harrison and Donnelly, 2011). Moreover, crowdsourcing initiatives leverage the collective intelligence of citizens to address urban challenges, from monitoring air quality to identifying traffic congestion patterns. This collaborative approach enhances the quality of public services and empowers residents to play an active role in shaping their city's future. However, successful citizen engagement in smart cities also necessitates addressing digital literacy disparities, ensuring that diverse voices are heard, and safeguarding data privacy to build trust between citizens and city authorities (Albino et al., 2015). Figure 14.7: Illustration of citizen engagement in smart cities.

14.7.1 Digital Platforms for Citizen Participation

Digital platforms for citizen participation are at the forefront of smart city initiatives, providing avenues for residents to engage with urban authorities and influence decision-making processes. These platforms encompass a wide range of tools, including mobile apps, websites, and social media platforms, enabling citizens to provide feedback, report issues, and participate in public consultations (Gupta et al.,

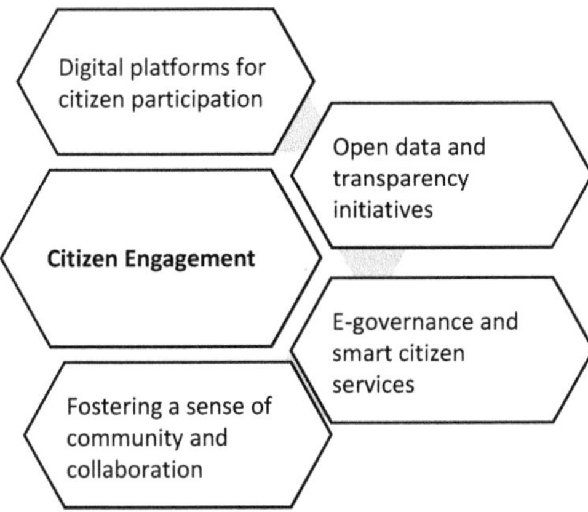

FIGURE 14.7 Fostering citizen engagement in the smart city ecosystem.

2017). Through these channels, residents can voice their concerns, contribute to urban planning, and monitor the progress of city projects, fostering transparency and inclusivity. However, for a fair and secure engagement of citizens in shaping the future of smart cities, challenges like the digital divide and concerns about data privacy need to be tackled.

14.7.2 Open Data and Transparency Initiatives

Open data and transparency initiatives are integral components of citizen engagement in smart cities, providing access to government data and decision-making processes. By making data related to urban services, budgets and planning available to the public in accessible formats, cities promote transparency, accountability and informed decision-making (Chourabi et al., 2012). Open data portals and platforms allow citizens, businesses and researchers to analyse and utilize data for various purposes, including urban planning, innovation and addressing civic challenges. However, ensuring data quality, addressing privacy concerns and fostering data literacy among citizens are crucial to realizing the full probability of exposed data and transparency ingenuities in smart cities.

14.7.3 E-governance and Smart Citizen Services

E-governance and smart citizen services play a crucial role in boosting citizen engagement and streamlining the delivery of public services within smart cities. Through digital platforms and applications, citizens gain easy access to government services, can submit requests, and conveniently track the progress of their inquiries. Moreover, e-governance initiatives streamline administrative processes, reduce bureaucracy, and enhance the responsiveness of government agencies, ultimately improving the overall quality of urban services (Moon, 2002). However, ensuring accessibility for all citizens, addressing digital literacy disparities, and safeguarding data security are crucial aspects that require careful attention in the implementation of e-governance and smart citizen services.

14.7.4 Fostering a Sense of Community and Collaboration

Fostering a sense of community and collaboration is a key goal in smart cities, enabled by digital platforms and initiatives that encourage residents to connect, share and work together for the betterment of their urban environment. Social media, community forums and neighbourhood apps facilitate interactions among residents, allowing them to organize events, discuss local issues and collectively address challenges. By promoting citizen-driven initiatives and participatory governance, smart cities empower residents to take an active role in shaping their communities, building trust and strengthening social cohesion. Nonetheless, ensuring that digital tools are inclusive, accessible and user-friendly for diverse populations is crucial to fostering a sense of community and collaboration in smart cities (Caragliu et al., 2011).

14.8 CHALLENGES AND CONSIDERATIONS

The growth and execution of smart city creativities are accompanied by a range of challenges and considerations that must be addressed to ensure their success and sustainability. These challenges encompass various dimensions, including technological, social and governance aspects. From a technological perspective, issues related to data privacy, security and interoperability pose significant hurdles. Ensuring the protection of citizens' personal data while enabling data sharing for urban analytics requires careful planning and robust cybersecurity measures. Additionally, the digital divide, unequal access to technology and disparities in digital literacy may exacerbate social inequalities, necessitating efforts to bridge these gaps (Caragliu et al., 2011). In terms of governance, the complexity of multi-stakeholder collaboration, the need for regulatory frameworks and the challenge of balancing innovation with responsible governance are paramount (Albino et al., 2015). Furthermore, the substantial financial investments required for smart city projects demand strategic planning and resource allocation (Albino et al., 2015). Addressing these challenges and considerations is essential to harness the full potential of smart city skills and confirm that the assistances are equitably distributed among urban populations.

14.8.1 Data Privacy and Security Concerns

Data privacy and security concerns represent critical encounters in the development and operation of smart cities, where vast amounts of personal and sensitive data are collected, processed, and shared. Protecting citizens' privacy while ensuring the secure handling of data is paramount. Unauthorized access, data breaches, and misuse of personal information pose significant risks, demanding robust cybersecurity measures, encryption protocols, and stringent access controls. Additionally, the ethical implications of data collection and surveillance must be addressed, alongside the establishment of clear regulations and compliance mechanisms to safeguard citizens' rights (Caragliu et al., 2011). Maintaining a delicate balance between reaping the benefits of data-driven urban management and safeguarding individual privacy requires continual attention and vigilance.

14.8.2 DIGITAL DIVIDE AND ACCESSIBILITY ISSUES

The digital gap and accessibility issues are weighty challenges in the progress of smart cities, as the equitable distribution of technological benefits among urban populations is crucial. Disparities in access to digital infrastructure, affordable internet connectivity, and technology literacy can exacerbate social inequalities, excluding certain groups from the benefits of smart city creativities (Caragliu et al., 2011). Bridging the digital divide requires targeted policies, public-private partnerships, and community-based initiatives to ensure that all residents have the opportunity to access and use smart city services. Furthermore, designing inclusive technologies and user-friendly interfaces is essential to enhance accessibility for individuals with disabilities and elderly populations, fostering a more inclusive and connected urban environment (Albino et al., 2015).

14.8.3 INTEROPERABILITY AND SCALABILITY CHALLENGES

Interoperability and scalability challenges are critical considerations in the implementation of smart city technologies, where diverse systems and devices must seamlessly connect and scale to meet evolving urban needs. Ensuring that data and technologies from various domains, such as transportation, energy, and healthcare, can communicate and collaborate effectively is essential for integrated smart city solutions. Achieving interoperability requires standardized protocols, open data formats, and well-defined interfaces. Scalability is equally crucial, as smart city infrastructure must be capable of accommodating growing populations and expanding services over time (Caragliu et al., 2011). Addressing these challenges necessitates robust governance frameworks, collaboration between stakeholders, and investments in technology infrastructure that can adapt and evolve as urban environments change.

14.8.4 BALANCING TECHNOLOGICAL INNOVATION WITH SOCIAL EQUITY

Balancing technological innovation with social equity is a fundamental challenge in the development of smart cities, where cutting-edge technologies often outpace considerations of their impact on vulnerable or underserved populations. The pursuit of innovation can inadvertently exacerbate existing disparities in access to resources, services, and opportunities, potentially leaving marginalized communities behind (Albino et al., 2015). Achieving social equity in smart cities requires deliberate efforts to ensure that technological advancements benefit all residents, regardless of socioeconomic status or background. This entails comprehensive urban planning, inclusive policy frameworks, and community engagement strategies that prioritize equitable access to technology, education, and opportunities for all citizens (Caragliu et al., 2011). Striking a balance between these priorities is vital to fashion smart cities that embody not just technological advancement but also social justice and inclusivity.

FIGURE 14.8   Unveiling the benefits and gazing into the future of smart cities (Kessler, 2019).

14.9   BENEFITS AND FUTURE OUTLOOK

The benefits of smart cities are multifaceted and hold immense promise for the future of urban living. These benefits encompass improved resource management, enhanced efficiency of urban services, reduced environmental impact, and elevated quality of life for residents. Smart cities optimize transportation networks, conserve energy and promote sustainable practices, contributing to reduced congestion, pollution, and energy consumption (Caragliu et al., 2011). They also advance healthcare access, public safety and citizen engagement, creating safer, more connected and inclusive urban environments (Albino et al., 2015). Peering into the future, the prospects for smart cities are filled with excitement, as progress in artificial intelligence, data analytics, and the Internet of Things stands ready to elevate the experience of urban living (Albino et al., 2015). However, realizing these benefits and ensuring their equitable distribution will require continued innovation, thoughtful governance, and a commitment to addressing challenges related to data privacy, accessibility, and social equity. As smart cities continue to evolve, they have the potential to reshape urban landscapes, improve the well-being of their inhabitants, and serve as models of sustainable and inclusive urban development. Figure 14.8: Overview of the benefits and future outlook of smart cities, focusing on advancements and the potential impact on urban living and sustainability.

14.9.1   Environmental Sustainability

Environmental sustainability is a cornerstone of smart cities, offering a range of benefits for the urban environment and its residents. By implementing innovative technologies and data-driven strategies, smart cities can optimize resource utilization, reduce carbon emissions and minimize their ecological footprint (Caragliu et al., 2011). Smart transportation solutions, for instance, promote sustainable mobility through efficient public transit systems, electric vehicles, and shared mobility services, ultimately mitigating traffic congestion and air pollution (Zheng et al., 2019). Moreover, think of smart energy management systems as the wizards behind the scenes, making our energy choices more Earth-friendly. They seamlessly weave renewable energy sources into the mix, whispering tales of reduced reliance on fossil fuels and an embrace of energy efficiency. It's like having a friend nudging us towards a greener, more sustainable way of powering our lives (Capozzoli et al., 2021). Smart waste management, with the aid of IoT-enabled bins and data analytics, minimizes waste generation and promotes recycling, contributing to cleaner, greener urban spaces. Overall, environmental sustainability in smart cities not only ensures a healthier, more eco-conscious urban lifestyle but also brings into line with worldwide exertions to combat climate variation and reserve natural resources (Albino et al., 2015).

14.9.2 ECONOMIC GROWTH AND INNOVATION

Economic growth and innovation are key outcomes of smart cities, propelling urban areas towards increased prosperity and competitiveness on a global scale. By leveraging advanced technologies and data-driven approaches, smart cities stimulate business development, entrepreneurship, and job creation (Caragliu et al., 2011). In the realm of smart transportation, envision a harmony orchestrated by autonomous vehicles and clever traffic management. Together, they fine-tune mobility, trim travel times, and elevate the entire ballet of urban logistics. This isn't just about efficiency; it's a melody that resonates with economic productivity, transforming the way we navigate our cities and making every journey a rhythm in the dance of urban life (Zheng et al., 2019). Furthermore, smart energy management and the integration of renewable energy sources foster energy efficiency, reducing operational costs for businesses and residents alike (Capozzoli et al., 2021). The efficient use of resources and improved infrastructure in smart cities attract investment, spur innovation and support the growth of knowledge-based industries, contributing to long-term economic sustainability (Albino et al., 2015). Thus, smart cities represent hubs of economic activity, where innovation and growth are nurtured in tandem with improved quality of life.

14.9.3 IMPROVED QUALITY OF LIFE FOR CITIZENS

Smart cities hold the promise of significantly refining the eminence of life for their inhabitants through enhanced urban services, efficient resource management and improved overall well-being (Caragliu et al., 2011). The implementation of smart transportation systems reduces traffic congestion, lowers commute times and minimizes the environmental impact of transportation, leading to less stressful and more pleasant daily journeys for citizens (Zheng et al., 2019). Additionally, smart healthcare systems, including telemedicine and remote patient monitoring, provide better access to medical services and empower citizens to take control of their health, resulting in improved health outcomes (Kouroubali et al., 2019). Moreover, smart cities prioritize public safety through data-driven law enforcement and emergency response, fostering safer environments and greater peace of mind for residents. The integration of digital platforms for citizen engagement encourages participation and collaboration, giving residents a voice in urban decision-making and strengthening their sense of belonging and community (Albino et al., 2015). These improvements collectively contribute to a higher eminence of life and comfort for citizens, making smart cities a model for the urban environments of the future.

14.9.4 Potential Future Advancements in Smart City Technologies

Peering into the horizon of smart cities reveals a tapestry of thrilling possibilities, as the march of progress continues with advancements in technologies like artificial intelligence, data analytics, and the Internet of Things (IoT). These innovations are poised to further transform urban living by enabling real-time data-driven decision-making, autonomous systems, then enhanced resident services (Albino et al., 2015). As AI and machine learning algorithms become more sophisticated, they can optimize traffic management, predicting traffic patterns and dynamically adjusting transportation routes (Zheng et al., 2019). The IoT will continue to expand its reach, connecting an ever-growing number of devices and sensors, facilitating smart waste management, and enabling precise energy consumption monitoring (Zheng et al., 2019). Moreover, smart cities are expected to evolve into interconnected ecosystems, fostering seamless integration between various sectors and services, ultimately creating more effective, bearable and liveable urban environments (Caragliu et al., 2011). However, these advancements also raise concerns related to data privacy, security and social equity, necessitating a careful and inclusive approach to future smart city development. Nonetheless, the potential for improved quality of life and urban sustainability through continued technological innovation makes the future of smart cities an exciting and transformative prospect.

14.10 CONCLUSION

In conclusion, the concept of smart cities and urbanization systems represents a transformative approach to addressing the complex challenges of rapid urbanization. By harnessing advanced technologies, data analytics, and interconnected infrastructure, smart cities aim to make additional bearable, efficient, and liveable urban atmospheres. This in-depth exploration has ventured into diverse facets of smart cities, spanning smart transportation, energy management, waste control, healthcare, public safety and citizen engagement. While the adoption of these technologies brings substantial potential benefits, it's imperative to acknowledge and address challenges such as data privacy and security, the digital divide, interoperability issues, and the imperative to harmonize technological innovation with social equity. Nevertheless, the benefits of smart cities are substantial, including environmental sustainability, economic growth, improved excellence of life for residents, and the potential for future advancements that will continue to shape the urban landscape. As smart cities continue to evolve, they hold the promise of becoming the model for urban living, driving innovation, and fostering more connected, prosperous and sustainable urban futures.

REFERENCES

Albino, Vito, Berardi, Umberto, & Dangelico, Rosa Maria. (2015). "Smart cities: Definitions, dimensions, performance, and initiatives." *Journal of Urban Technology*, 22(1), 3–21.

Al-Hader, M., Thakuriah, P., & Ozbay, K. (2016). Smart Surveillance for Smart Cities: A review. *Journal of Intelligent Transportation Systems*, 20(3), 260–285. doi:10.1080/15472450.2015.1119620

Arena, U., Ardolino, F., Di Gregorio, F., & Iacobucci, R. (2019). Smart waste collection and transportation systems: A review. *Waste Management*, 87, 41–58.

Balderrama-Subieta, S., Difiglio, C., Cabezas, L. Á., & Pereira, M. G. (2017). Urban energy consumption and greenhouse gas emissions in Latin America: Exploring the role of urban form in medium and large cities. *Energy Policy*, 105, 312–326.

Bieńkowski, A., Gomółka, J., & Kijewska, K. (2018). Multimodal transportation in smart cities: A review. *Sustainable Cities and Society*, 38, 789–797.

Bogar, D. S. (2017). Policing by the numbers: Big data and the Fourth Amendment. *Washington Law Review*, 92(4), 959–1026.

Capozzoli, A., Ciribini, A. L. C., & Riva Sanseverino, E. (2021). "Smart technologies and urban energy management: A literature review". *Energies*, 14(2), 425.

Caragliu, A., Del Bo, C., & Nijkamp, P. (2011). "Smart cities in Europe". *Journal of Urban Technology*, 18(2), 65–82.

Chen, J., Zhang, D., Wang, S., & Cai, W. (2018). Data mining for the internet of things: Literature review and challenges. *International Journal of Distributed Sensor Networks*, 14(12), 1550147718814881. doi:10.1177/1550147718814881

Chien, S., & Ding, C. (2019). Toward autonomous vehicles: Review of the recent patents in the field. *Recent Patents on Electrical & Electronic Engineering*, 12(2), 94–106.

Chourabi, H., Nam, T., Walker, S., Gil-Garcia, J. R., Mellouli, S., Nahon, K., … & Scholl, H. J. (2012). Understanding smart cities: An integrative framework. In *2012 45th Hawaii International Conference on System Sciences* (pp. 2289–2297). IEEE. doi:10.1109/HICSS.2012.615

Ding, C., Wei, Y., Ma, S., & Ding, H. (2019). A review of urban intelligent transportation systems: Concepts, evolution, and challenges. *Transportation Research Part C: Emerging Technologies*, 98, 63–76.

Dutta, A., Anisimov, E., & Sircar, S. (2019). Renewable energy in smart cities: Challenges and opportunities. *Energy Procedia*, 156, 74–79.

Geng, Y., Fu, J., Sarkis, J., Xue, B., & Fujita, T. (2019). Towards a national circular economy indicator system in China: An evaluation and critical analysis. *Journal of Cleaner Production*, 215, 775–785.

Gupta, S., Gupta, R., & Singla, C. (Aug. 2017). "Analysis of image enhancement techniques for astrocytoma MRI images," *International Journal of Information Technology*, 9(3), 311–319. doi: 10.1007/s41870-017-0033-8

Gupta, S., & Singla, C. (2016). Grade identification of astrocytoma using image processing — A literature review. In *2016 3rd International Conference on Computing for Sustainable Global Development* (pp. 1968–1973). INDIACom.

Hao, H., Amin, S., & Li, F. (2019). The role of demand response in smart grid operations: A review. *IEEE Transactions on Industrial Informatics*, 15(6), 3435–3444.

Harrison, C., & Donnelly, I. A. (2011). "A theory of smart cities", *Proceedings of the 55th Annual Meeting of the ISSS - 2011*, 169–181.

Inxee Systems Private. (2023). Cloud Computing for Smart Cities: Advantages & Features, https://www.linkedin.com/pulse/cloud-computing-smart-cities-advantages-features-inxee-systems/

IoT Misr. (2020). Smart Cities, Smart waste management. https://iot-misr.com/solutions/smart-cities/

Jin, W., Zhang, L., & Ding, Y. (2016). A review of dynamic traffic management strategies. *Transportation Research Part C: Emerging Technologies*, 67, 168–179.

Joy, P. (October 8, 2022). Smart Grid Enables Quantum Improvement in Energy Management Efficiency, https://www.powerelectronicsnews.com/smart-grid-enables-quantum-improvement-in-energy-management-efficiency/

Kambatla, K., Barat, C., & Liu, H. (2019). Emerging challenges in cybersecurity research for connected autonomous vehicles. *ACM Computing Surveys (CSUR)*, 52(2), 1–38.

Keates, S. (2019). Accessibility issues for connected and autonomous vehicles: A review. *Applied Sciences*, 9(5), 1027.

Kessler, Dietmar. (2019). Future Smart City – How the Internet of Things is Transforming Our Cities, https://www.dotmagazine.online/issues/digital-infrastructure-foundation/the-internet-of-the-future/future-smart-city

Korhonen, J., Honkasalo, A., & Seppälä, J. (2018). Circular economy: The concept and its limitations. *Ecological Economics*, 143, 37–46.

Kouroubali, A., Roumeliotis, S., & Maramis, C. (2019). "Smart cities and data-driven policy", *Journal of the American Medical Informatics Association*, 26(12), 1674–1679.

Li, W., Shafie-khah, M., Catalão, J. P., & Xia, Q. (2013). A review of the single and combined impacts of wind and solar power generation on the power system. *Renewable and Sustainable Energy Reviews*, 25, 381–396.

Li, X., Li, X., Huang, H., Xie, X., & Liu, W. (2017). A survey of urban computing for smart cities. *Information Fusion*, 38, 74–89.

Litman, T. (2018). *Autonomous vehicle implementation predictions: Implications for transport planning*. Victoria Transport Policy Institute.

Litman, T. (2019). *Congestion Costing Methods and Results*. Victoria Transport Policy Institute.

Litman, T. (2020). *Evaluating Transportation Equity: Guidance for Incorporating Distributional Impacts in Transportation Planning*. Victoria Transport Policy Institute.

Lu, S., Zhang, G., Wang, J., Li, Y., & Dong, L. (2017). Demand response in the smart grid: A review. *IEEE Transactions on Industrial Informatics*, 13(6), 3022–3031.

Mandl, K. D., Gottlieb, D., Ellis, A., & Cullen, R. (2020). Smart Health Community: The Hidden Value of Health Data. *New England Journal of Medicine*, 382(23), 2285–2287. doi:10.1056/NEJMp2000315

Mohler, G., Short, M. B., Malinowski, S., Johnson, M., Tita, G. E., Bertozzi, A. L., & Porter, J. R. (2015). Randomized controlled field trials of predictive policing. *Journal of the American Statistical Association*, 110(512), 1399–1411. doi:10.1080/01621459.2015.1077710

Moon, M. J. (2002). The evolution of e-government among municipalities: Rhetoric or reality? *Public Administration Review*, 62(4), 424–433.

Nam, T., & Tan, Y. (2016). "Smart city as urban innovation: A reflective analysis", *Information Systems Frontiers*, 18(2), 263–286.

Orozco-Arroyave, J. R., Arias-Arroyo, A., Ríos-Ortega, D., Lopéz-Gamba, J., & Álvarez-Meza, A. M. (2018). Smart cities and their cybersecurity and privacy challenges: A comprehensive review. *IEEE Access*, 6, 19656–19676. doi:10.1109/ACCESS.2018.2807781

Pandey, A. K., Panigrahi, J. K., & Singh, A. (2016). Solar energy for the future world: A review. *Renewable and Sustainable Energy Reviews*, 62, 1092–1105.

Poongodi, M., Sharma, A., Hamdi, M. et al. (2021). Smart healthcare in smart cities: wireless patient monitoring system using IoT. *J Supercomput*, **77**, 12230–12255. https://doi.org/10.1007/s11227-021-03765-w

Pulido, J., Kämpf, J., & Axhausen, K. W. (2019). Predictive maintenance in urban public transport: Identifying maintenance needs in tram infrastructure using sensor data. *Transportation Research Part C: Emerging Technologies*, 98, 133–150.

Sivakumar, A., Schoettle, B., & Sadek, A. W. (2018). Utilizing emerging data sources to gain insights into bicycle and pedestrian travel behavior: A case study in Ann Arbor, Michigan. *Transportation Research Part C: Emerging Technologies*, 95, 98–117.

Steve Mazur, Business Development Director, Government, December 09, 2020, https://www.digi.com/blog/post/introduction-to-smart-transportation-benefits

Tukker, A., Aurich, P., Emmanuel-Yusuf, D., Lutter, S., Martinho, G., & Tischner, U. (2019). Environmental and resource footprint assessment (ERFA) of household consumption in the EU27: A cross-country time-series analysis from 2000 to 2013. *Environmental Research Letters*, 14(6), 065006.

Tuli, S., Tuli, S., & Samuel, A. (2020). Smart hospitals: An IoT-based approach for patient health monitoring and management. In *Handbook of Healthcare in the IoT and Blockchain* (pp. 1–22). Elsevier. doi:10.1016/B978-0-12-818890-9.00001-1

United Nations. (2018). *68% of the world population projected to live in urban areas by 2050, says UN.* Retrieved from https://www.un.org/development/desa/en/news/population/2018-revision-of-world-urbanization-prospects.html

Vegesna, A., Tran, M., Angelaccio, M., & Arcona, S. (2017). Remote Patient Monitoring via Non-Invasive Digital Technologies: A Systematic Review. *Telemedicine and e-Health,* 23(1), 3–17.

Wai Yip St. (2022). viAct for Smart Cities, Kwun Tong, Hong Kong, https://www.viact.ai/smartcities

Xie, K., Guo, X., Wang, Z., Ma, L., & Lin, H. (2017). A review of smart parking systems. *IEEE Access,* 5, 28718–28731.

Zhang, H., Zheng, Y., & Zhao, L. (2018). "Understanding urban dynamics from inter-city movement data", *Proceedings of the 24th ACM SIGKDD International Conference on Knowledge Discovery & Data Mining,* 2453–2461.

Zheng, Y., Li, L., Chen, J., Xie, X., & Ma, W. Y. (2019). "Understanding transportation modes based on GPS data for web applications", *ACM Transactions on the Web (TWEB),* 13(1), 1–28.

15 Empowering the Future
Digital Twins and Connectivity in Smart City Development

*Abdallah Abualkishik, Marwan Alshar'e,
Ahmad Kayed, Khaled Abuhmaidan and
Ala Odeibat*
Sohar University, Sohar, Oman

15.1 INTRODUCTION

A "smart city" is a concept that involves using advanced technologies and data-driven strategies to improve the quality of life, sustainability and efficiency of urban areas (Winkowska, Szpilko, and Pejić 2019). In the late 1990s, several cities began implementing smart initiatives. One notable example is the city of Barcelona, which launched its program in 2011 (Mora, Bolici, and Deakin 2017). Barcelona focused on integrating ICT solutions in areas such as energy management, mobility and public services. The success of Barcelona's initiatives inspired other cities worldwide to adopt similar approaches. These smart initiatives aim to improve citizen quality of life, optimize city operations and foster economic growth. Instead of just focusing on how much technologies are available, the value is in how to integrate these technologies to work together in efficient and green manner, with respect for other diminutions (such as cost, time, human resource and others).

In (Lee et al. 2018), the utilization of Fourth Industrial Revolution technologies in the process and energy industries is anticipated to bring about advancements in energy management and optimization, enhanced servicing and maintenance, the development of energy-efficient designs, and the evolution of current sites. Additionally, the integration of locally and regionally generated renewable energy is expected to be a key outcome of this digitalization initiative. The fourth industrial revolution, driven by advancements in artificial intelligence, has taken the digital revolution to a higher level. It has led to the emergence of "Globalization 4.0," encompassing economic, political, and cultural globalization, as announced by the World Economic Forum (Surya and Fikriya 2021). The historical trajectory of smart cities reveals a dynamic evolution, shaped by the convergence of technological advancements and urban development.

Early precursors to the smart city concept date back to the latter half of the 20th century, with the vision of leveraging technology to enhance urban living (Albino,

DOI: 10.1201/9781003467892-15

Berardi, and Dangelico 2015). As cities evolve into complex ecosystems, the integration of cutting-edge technologies becomes imperative for addressing the multifaceted challenges that accompany rapid urbanization. Scholars have long recognized the transformative potential of smart cities in elevating urban living standards, fostering sustainability and optimizing operational efficiency within the intricate fabric of urban areas (Park, Del Pobil, and Kwon 2018).

Artificial Intelligence (AI) has emerged as a transformative force with the potential to transform various aspects of our lives, and its role in the design of smart cities. Currently, AI algorithms and applications are involved in all real-life aspects (Lee and Yoon 2021). AI powers the predictive maintenance in manufacturing to optimize operations and reduce downtime and provides intelligent chatbots for customer service to enable small businesses to offer 24/7 support without extensive resources. AI enables e-commerce platforms in emerging markets, boosting small business visibility and sales. Moreover in (Elliott 2019), AI drives agricultural technologies to optimize crop yields and support agribusiness diversification, AI helps households and industries to analyze satellite imagery for monitoring deforestation and habitat preservation and optimizes consumption using Grid Computing technology.

In waste management systems, (Benzidia, Makaoui, and Bentahar 2021), and (Ghoreishi and Happonen 2020) emphasize that AI improves the collection routes, reducing energy consumption and emissions, predicts demand, and minimizes waste. AI has the power to analyze preferences and behaviours to create personalized travel recommendations and itineraries, facilitate seamless communication between tourists and locals, break language barriers, offer immersive experiences, and allow people to explore cultural landmarks and historical sites virtually in the tourism sectors. AI also enhanced safety measures by monitoring and analyzing crowds, detecting potential risks, and ensuring a safe environment for humanity (Alhayani et al. 2021). (Huang, Saleh, and Liu 2021) stated that AI plays a key role in personalizing learning platforms that adapt to each student's pace and learning style, enhances skill assessment platforms that guide individuals in choosing the most relevant skills to develop, and provides a well-established platform to analyze a person's skills and aspirations to recommend suitable career paths and educational opportunities.

Efficiency and reliability of connectivity play a critical role in the success of smart cities. The emergence of the Internet of Things (IoT) and Big Data Analytics has played a pivotal role in catalyzing the realization of these visionary concepts. IoT, with its network of interconnected devices, has enabled the seamless flow of data, forming the foundation for smart city ecosystems (Alharbi and Soh 2019). Simultaneously, Big Data Analytics has emerged as a powerful tool for extracting meaningful insights from the vast datasets generated within urban environments. These technologies collectively serve as the linchpin for the development of smart cities, facilitating data-driven decision-making and innovation. The proliferation of the Internet of Things (IoT) and big data analytics further accelerated the development of smart cities. The ability to collect and analyze vast amounts of data from sensors and devices enabled cities to gain valuable insights for better decision-making and resource management, (Telo 2023). As we delve into the heart of this exploration, it becomes evident that the discourse on smart cities extends beyond mere technological adoption. It encompasses a vision for urban development that transcends traditional paradigms, acknowledging the role of technology as a catalyst for positive change. Through an interdisciplinary lens, it is essential to acknowledge the diverse perspectives that inform the discourse on smart cities. Scholars from urban planning, computer science, sociology and environmental studies converge to explore the multifaceted implications of smart city initiatives (Karimi et al. 2021).

By critically examining the historical evolution and the role of key technologies, this paper seeks to provide a nuanced understanding of the potential and challenges inherent in the smart city paradigm, aspires to offer a holistic examination of smart cities that transcends disciplinary boundaries, and delves deeper into the advantages and challenges posed by smart cities, exploring how they impact various facets of urban life. The focus will extend also to the transformative role of digital twins and the fundamental significance of connectivity in smart city operations. Through a synthesis of academic research, case studies and practical applications, this paper moreover seeks to provide a comprehensive understanding of the intricate dynamics shaping the future of smart cities.

15.2   ADVANTAGES AND CHALLENGES OF SMART CITIES

Smart cities represent a revolution in urban development, using digital technologies to increase efficiency, sustainability and overall quality of life for citizens. One outstanding benefit is the improving infrastructure. Smart cities use advanced sensors and data analytics to optimize energy consumption, waste management and water distribution, reduce environmental impact and provide infrastructure well increases Furthermore, real-time analytics by integrating Internet of Things (IoT) devices, Data-driven decision-making is also made easier, enabling authorities to tackle emerging challenges such as traffic quickly responding to congestion or environmental hazards this in turn contributes to increased safety and resilience (Ahad et al. 2020). In addition to resource efficiency, smart cities support economic growth through increased innovation and connectivity. The use of smart infrastructure, including smart transportation systems and high-speed connectivity, attracts jobs and talent, creating an enabling environment for economic growth. For example, the use of telemedicine and intelligent educational systems improves the quality and scale of essential services. Furthermore, the establishment of an integrated ecosystem facilitates the development of innovative solutions and employment, which generates economic prosperity (Abuhaija et al. 2023) & (Bibri and Krogstie 2020).

Many researchers as Khan et al. (2020), Trindade et al. (2017), and Huang, Kim, and Schermer (2022) mentioned that while the benefits of smart cities are compelling, their development presents a number of challenges that need to be carefully considered. Privacy and security concerns stand out as major obstacles. The massive amount of data collected by smart city systems, including personal data and real-time location data, raises privacy concerns. Ensuring strong cybersecurity measures is essential to protect citizens' data and prevent unauthorized access that could compromise both individual privacy and the integrity of the smart city initiative. Another challenge revolves around the digital divide, where not all citizens share the benefits of smart technologies equally. Socioeconomic inequalities can lead to unequal access to digital services, exacerbating existing inequalities. To address this, smart city policies must be inclusive, ensuring that technological advances benefit all segments of society Furthermore, the integration of new technologies may face resistance from citizens worried about job displacement or cultural change. Effective communication and community engagement are essential to building trust and gaining support for smart city initiatives. The development of smart cities presents a transformational approach to urban life, offering many benefits in terms of resource efficiency, economic growth and improved public services but overcoming challenges of privacy, security, addressing inclusion and social recognition.

15.3 DIGITAL TWINS AND SMART CITY MANAGEMENT

In the field of smart city management, (Israilidis et al. 2021) highlighted that the integration of sophisticated technologies has become essential to create more efficient, sustainable, and responsive urban environments. Among these transformative technologies, Building Information Modelling (BIM) stands out as a foundational concept that has revolutionized the way we conceptualize and construct physical structures. BIM facilitates a comprehensive digital representation of buildings, allowing for accurate planning and collaborative decision-making throughout the construction process (Xia et al. 2022). From another perspective, (Luusua et al. 2023) pointed out that integrating AI into smart cities is critical to addressing the raised challenges of urban living. By utilizing the capabilities of AI, smart cities can be sustainable, efficient, and liveable places for their residents. The ongoing development and application of AI technologies will shape the future of urban life, encouraging innovation and improving overall quality of life.

4th industrial revolution, AI technologies, the concepts of information-rich modelling like BIM, with respect for security and ethical matters leads scholars forward to set the stage for an even more advanced innovation on these technologies. However, researchers Ahmad et al. (2022), Aigbavboa et al. (2023), and Jiang et al. (2020) from different fields stated that this technology still facing great challenges. Existing infrastructure in cities may not be compatible or equipped to support the integration of smart technologies. Retrofitting or upgrading infrastructure to accommodate sensors, connectivity and data management systems can be a complex and costly process (Nkwunonwo et al. 2023). Smart cities involve multiple systems, devices and platforms that need to communicate and interoperate seamlessly. Lack of interoperability and standardization across different technologies and vendors can hinder integration efforts and limit the scalability and effectiveness of smart city solutions (Jafari et al. 2023). Implementing smart city initiatives often requires coordination among various government departments, agencies and stakeholders. Overcoming bureaucratic barriers, establishing clear governance structures, and facilitating cross-departmental collaboration can be challenging.

Developing robust data governance frameworks, complying with privacy regulations, and addressing concerns related to data ownership and consent are essential for successful implementation (Wirtz and Müller 2023). Therefore, there is a need to narrow down these challenges and implement up-to-date technologies to facilitate smart cities implementation. Researchers and scholars started a few years ago to reevaluating an old concept which was presented 50 years ago by NASA which was the spark of Digital Twin (Grieves 2023). NASA built an adaptable simulation to match conditions of crippled spacecraft to run and reject different scenarios. Despite that, the simulation was limited and not intelligent. In 2003, (Qu et al. 2023) Grieves defined the three parts: the physical entity, the digital entity and connection between them. NASA back in 2010 to define the formal name for the concept Digital Twin through their space vehicle digital layer. Grieves, in 2014, expanded the application of the Digital Twin concept to industries such as Gas and Oil, automation, and healthcare. Today's creation of smart cities is thought to have a new beginning in the form of digital twins (Grieves 2023) & (Qu et al. 2023).

Digital twin technology serves as a virtual model for processes, products, and services. It provides companies and industries with valuable insights for data analysis, system monitoring, proactive problem-solving, prevention of breakdowns, creation of new opportunities, and preparation for the future through simulations (Shahat, Hyun, and Yeom 2021). Digital Twins takes the principles of BIM to the next level by providing dynamic virtual replicas of physical objects or systems. These replicas are not static; They are developed in real time, integrating data from a variety of sources. Digital Twins are powerful tools in the context of smart city management, providing a dynamic and responsive platform for informed decision-making (Sadri et al. 2023). Digital Twins became increasingly integrated with the 4th Industry revolution pillars such as AI, IoT, Big-data analytic, Cloud & Grid computing, Blockchain, 5G Network, Augmented & Virtual reality, and Robotics & Automation technologies. With respect to security and ethics perspective, the integration enhances the ability of Digital Twins to gather real-time data from sensors and make intelligent predictions about the physical assets they represent.

Figure 15.1 Conceptual Abstraction Model for Digital Twin that clarifies the integration of three main layers. First layer represents the 4th industrial pillar as a base for smart City development. The second layer is called Building Information Modelling (BIM), which is a real-time digital representation using modelling, visualization and simulation of a building's or infrastructure's functional and physical features. In addition, the Digital Twin layer is as interactive real-time response and management layer using the benefit of the first two layers.

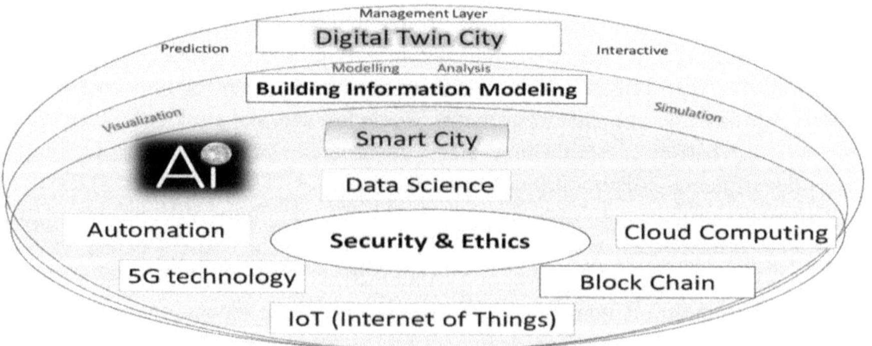

FIGURE 15.1   Conceptual abstraction model for Digital Twin.

Digital twins distinguish themselves from Building Information Modeling (BIM) due to their encompassing nature. While BIM's primary goal is to generate precise virtual representations of assets, digital twins take it a step further by integrating real-time data and accurately reflecting the present condition of the physical asset. In Deng, Menassa, and Kamat (2021), authors stated that BIM is used more for operations and maintenance visualization in design and construction, in contrast, Digital Twin concentrates on how people interact with built environments. (Deng et al. 2021). As a management layer, Digital Twin layer is working as the brain for the smart city to make it green. It provides a real-time response to control, manage, and mitigate smart city working flow.

15.4 CONNECTIVITY AS A FUNDAMENTAL ENABLER

Connectivity solutions have become essential for people worldwide, enabling them to stay connected, productive and informed (Alharbi and Soh 2019). Governments around the world are currently racing to integrate technology into various aspects of city operations, including public transportation, IT connectivity, water and power supply, sanitation and waste management, urban mobility, e-governance, citizen participation, education, and more (Ahvenniemi et al. 2017). The fourth industrial revolution, driven by advancements in artificial intelligence, has taken the digital revolution to a higher level. It has led to the emergence of "Globalization 4.0," encompassing economic, political and cultural globalization, as announced by the World Economic Forum (Surya et al. 2021).

In the context of digital twin technology, a virtual model visualizes the physical smart city and generates real-time data through simulation technology. This simulation acts as a dynamic collaboration dashboard for managing, analyzing and tolerating all federated data within the city (Hatuka and Zur 2020). Efficiency and reliability of connectivity play a critical role in the success of smart cities. Connecting numerous nodes and transmitting vast amounts of data requires the implementation of the latest technologies. The concept of the Internet of Everything (IoE) has emerged to expand the functionality of the Internet of Things (IoT), enabling machine-to-machine (M2M), people-to-machine (P2M), and vice versa (M2P) communication. This technology generates massive real-time data transmission through wired or wireless connections. Researchers argue that cloud, edge and grid computing, combined with IoE, are essential for enhancing operational efficiency in smart cities (Alharbi and Soh 2019). In smart cities, underground broadband infrastructure is utilized for gas, electricity, and water utilities. Broadband networks have become the fourth utility that society heavily relies on. Metro broadband networks employ fibre networks for ultrafast connectivity, promoting existing and emerging technologies and facilitating digital transformation in smart cities. This enables real-time interaction between the physical and virtual models in the digital twin. Wireless connection quality and speed are also significant considerations for the digital twin.

The fifth-generation mobile network (5G) serves as an innovation platform, enabling mass connections over larger areas with low latency. These advantages result in improved and faster calculations for various applications, such as traffic management and emergency response (Alharbi and Soh 2019). The digital twin of smart cities has an application layer that serves four main areas: home, transport, energy, and health. Each of these areas requires different types of sensors, including electronic, chemical, smart-grid, and biosensors. To ensure seamless cooperation and data exchange, the multiple systems within these areas need to be linked across application, edge and cloud layers (Liu et al. 2019).

Connectivity is now considered a utility, especially for businesses that rely on providers and consumers of data. Broadband infrastructure, particularly fibre networks, supports the development of smart cities and enables digital transformation (Schroeder et al. 2021). Digital twin technology, along with advancements in connectivity and sensors, plays a crucial role in monitoring various aspects of daily living in smart cities, such as crowd density, cleanliness of public spaces and vehicle movement.

15.5 APPLICATION OF DIGITAL TWINS IN VARIOUS DOMAINS

Digital twins, with their capacity to replicate and interact with real-world entities, have revolutionized several domains by leveraging the advantages of collecting and analyzing data related to infrastructure, patterns, people, and their responses among various domains and systems.

The virtual replications of real systems or objects, or "digital twins," have found a broad variety of uses across numerous fields, emphasizing their significant power to improve performance, decision-making, and overall operational performance (Wagg et al. 2020). Within the domain of home automation, digital twins work as virtual representations of the features and operational aspects of smart houses. With the help of these twins, homeowners may keep an eye on and manage several different aspects in real time, including energy usage, home appliances, and security systems. In this field, interoperability is essential since it enables smooth cooperation and integration between various smart devices. Digital twins make it possible for data to be shared across diverse systems, enabling the creation of a more intelligent and integrated living environment through defined communication protocols (Zheng, van Essen, and Bernhaupt 2023).

One more sector in which digital twins are making substantial improvements is transportation. Digital twins of automobiles assist in design, testing and scheduled repairs in the vehicle manufacturing sector. Whenever real models are constructed for real, manufacturers can evaluate component effectiveness and simulate various driving cases to improve solutions (AlKishri, Abualkishik, and Al-Bahri 2022). Connectivity is essential in such instances considering that it empowers data transmission between the virtual and real worlds, facilitating that simulation knowledge is readily utilized to improve the safety and performance of real vehicles. Additionally, digital twins of transportation systems in smart cities help in the flow of traffic management, bottleneck reduction and improving the effectiveness of general urban mobility (Jafari et al. 2023).

Digital twins aid in the effective administration of power plants, renewable energy sources, and distribution networks in the energy sector. These digital twins enable real-time insights into the amount of energy produced, patterns of consumption, and asset longevity, supporting predictive maintenance and efficient resource utilization

(Gao and Huang 2023). Interconnection is critical for improving integrity between various energy infrastructures and allowing for implementing a better sustainable approach in terms of power generation and distribution. Moreover, it facilitates data flow across various components of the energy ecosystem. In the health care domain, Digital twins have been adopted for individualized medicine, patient monitoring, and treatment planning (Kaul et al. 2023). Virtual representations of organs or entire biological systems allow healthcare experts to model and analyze various treatment scenarios, ultimately leading to better outcomes. Manufacturers stakeholders also can monitor process performance, spot abnormalities and increase production efficiency by harnessing AI technologies and developing digital identical copies of manufacturing lines and equipment (Abuhaija et al. 2023). Real-time data gathering facilitates quick responses to equipment failures or breaks from normal operating conditions, minimizing downtime and ensuring continuous working efficiency with respect to security issues (Albazar et al. 2023).

## 15.6	CONNECTIVITY AND DIGITAL TWINS: CATALYSTS FOR TRANSFORMATION

Smart cities are viewed as catalysts for revolutionary change due to the interplay between connectivity and digital twins. Connectivity and digital twins combine to transform urban development and governance by solving the complex issues of modern living (Bonney et al. 2021).

Smart cities rely on connectivity, that provides the framework for reliable interaction between systems, sensors, and devices. The integration of digital twin technology, which generates virtual copies of real assets, infrastructure and even whole metropolitan ecosystems, is based on this interconnected network. Connectivity and digital twins work together to provide real-time data transmission and analysis, opening new opportunities for inhabitants, administrators, and urban planners (Mashaly 2021).

Digital twins are used in urban planning to collect data from a wide range of sources, such as smart infrastructure, Internet of Things sensors, and citizen input. When combined with virtual models, this data provides a full and dynamic picture of the current state of the city. With the use of this data, urban planners can evaluate the effects of suggested modifications, simulate and analyze various scenarios, and make well-informed decisions that maximize resource allocation and improve overall urban functionality, (Voigt et al. 2021). Furthermore, public services and safety are included in the connectivity-driven integration of digital twins. Digital twin real-time data can be used by emergency response systems to quickly assess and address hazardous circumstances. For example, the interconnectedness of digital twins and connectivity allows authorities to efficiently manage public places, divert traffic and coordinate emergency services in the event of a natural catastrophe or traffic crisis (Schroeder et al. 2021).

The benefits of this mutually beneficial partnership also encompass public participation. Urban environment shaping is an active process that citizens can engage in via connected gadgets and digital twins. Public access to real-time data on energy use, air quality, and public services can promote a feeling of community involvement. By enabling individuals to participate in the decision-making process, this openness and engagement develop an urban governance model that is more responsive and inclusive. The combination of digital twins and connectivity will be crucial in tackling urban issues as smart cities develop (Jiang et al. 2021). Digital twins become more accurate and responsive due to the real-time data interchange that connection enables, making them essential tools for decision-makers. The influence of connectivity on the capabilities of digital twins will be further amplified by the ongoing development of IoT infrastructure, 5G networks and edge computing. This will open the door to even more advanced applications in the areas of sustainability, urban development, and public welfare. Revolutionary wave in smart cities is being driven by the interaction between connectivity and digital twin technology. Incorporating virtualized urban landscapes with real-time data interchange empowers decision-makers and involves individuals in co-creating their living spaces. A new era of urban transformation is beginning as this symbiotic relationship continues to develop and the idea of truly intelligent and interconnected cities comes to pass (Caprari et al. 2022).

15.7 CONCLUSION

In the pursuit of realizing the vision of smart cities, this paper has delved into the transformative intersection of digital twins and connectivity, laying the groundwork for a future where urban development is synonymous with technological innovation. The concept of a smart city, driven by advanced technology and data-driven strategies, holds the promise of elevating living standards, fostering sustainability, and optimizing urban services. However, the practical implementation of smart cities poses challenges such as high costs, infrastructure compatibility, governance issues, and privacy concerns, necessitating innovative solutions for effective administration.

A significant breakthrough in addressing these challenges is the emerging concept of digital twins – dynamic virtual representations intricately linked with real-time data. These digital counterparts offer a sophisticated toolset for decision-makers, enabling dynamic and informed management of smart cities. This paper has explored the applications of digital twins in key domains including home automation, transportation, energy and healthcare, emphasizing the crucial role of interoperability for seamless collaboration and data exchange among diverse systems. Connectivity, identified as a fundamental utility for smart cities, takes centre stage as the linchpin for driving digital transformation. With a focus on broadband infrastructure and the advent of 5G technology, connectivity becomes the lifeblood that empowers the functionalities of digital twins. The intricate interplay between connectivity and digital twins facilitates real-time data exchange, enabling cities to respond swiftly to dynamic urban challenges.

As smart cities continue to evolve, embracing these technologies will be pivotal in overcoming existing barriers and unlocking the full potential of technology for the benefit of urban dwellers. With a shared vision of better living standards, resource management, and economic vitality, the marriage of digital twins and connectivity emerges as a transformative force, empowering the future of smart city development. The journey towards smart cities is not just a technological evolution but a collective endeavour to build communities that are resilient, innovative and equipped to meet the challenges of the future.

REFERENCES

Abuhaija, Belal, Aladeen Alloubani, Mohammad Almatari, Ghaith M. Jaradat, Barzan Abdallah Hemn, Abdallah Mohd Abualkishik, and Mutasem Khalil Alsmadi. 2023. "A comprehensive study of machine learning for predicting cardiovascular disease using Weka and SPSS tools." *International Journal of Electrical and Computer Engineering* 13(2): 1891.

Ahad, Mohd Abdul, Sara Paiva, Gautami Tripathi, and Noushaba Feroz. 2020. "Enabling technologies and sustainable smart cities." *Sustainable Cities and Society* 61: 102301.

Ahmad, Kashif, Majdi Maabreh, Mohamed Ghaly, Khalil Khan, Junaid Qadir, and Ala Al-Fuqaha. 2022. "Developing future human-centered smart cities: Critical analysis of smart city security, data management, and ethical challenges." *Computer Science Review* 43: 100452.

Ahvenniemi, Hannele, Aapo Huovila, Isabel Pinto-Seppä, and Miimu Airaksinen. 2017. "What are the differences between sustainable and smart cities?." *Cities* 60: 234–245.

Aigbavboa, Clinton, Andrew Ebekozien, and Nompumelelo Mkhize. 2023. "A qualitative approach to investigate governance challenges facing South African airlines in the fourth industrial revolution technologies era." *Social Responsibility Journal* 19, no. 8: 1507–1520.

Albazar, Hussein, Ahmed Abdel-Wahab, Marwan Alshar'e, and Abdallah Abualkishik. 2023. "An adaptive two-factor authentication scheme based on the usage of Schnorr Signcryption algorithm." *Informatica* 47: 159–172.

Albino, Vito, Umberto Berardi, and Rosa Maria Dangelico. 2015. "Smart cities: Definitions, dimensions, performance, and initiatives." *Journal of Urban Technology* 22(1): 3–21.

Alharbi, N., and B. Soh. 2019. "Roles and challenges of network sensors in smart cities." *In IOP Conference Series: Earth and Environmental Science* 322(1): 012002.

Alhayani, Bilal, Husam Jasim Mohammed, Ibrahim Zeghaiton Chaloob, and Jehan Saleh Ahmed. 2021. "Effectiveness of artificial intelligence techniques against cyber security risks apply of IT industry." *Materials Today: Proceedings*. 531.

AlKishri, Wasin, Abdallah Abualkishik and Mahmood Al-Bahri. 2022. "Enhanced image processing and fuzzy logic approach for optimizing driver drowsiness detection." *Applied Computational Intelligence and Soft Computing* 2022, no. 1 9551203: 1–14.

Benzidia, Smail, Naouel Makaoui, and Omar Bentahar. 2021. "The impact of big data analytics and artificial intelligence on green supply chain process integration and hospital environmental performance." *Technological Forecasting and Social Change* 165: 120557.

Bibri, Simon Elias, and John Krogstie. 2020. "The emerging data–driven Smart City and its innovative applied solutions for sustainability: The cases of London and Barcelona." *Energy Informatics* 3: 1–42.

Bonney, Matthew S., Paul Gardner, David Wagg, and Robin Mills. 2021. "Case study of connectivity of digital twins and experimental systems." *In Proceedings of the 8th International Conference on Computational Methods in Structural Dynamics and Earthquake Engineering*. Streamed from Athens, Greece: 1416–1425.

Caprari, Giorgio, Giordana Castelli, Marco Montuori, Marialucia Camardelli, and Roberto Malvezzi. 2022. "Digital twin for urban planning in the green deal era: A state of the art and future perspectives." *Sustainability* 14(10): 6263.

Deng, Min, Carol C. Menassa, and Vineet R. Kamat. 2021. "From BIM to digital twins: A systematic review of the evolution of intelligent building representations in the AEC-FM industry." *Journal of Information Technology in Construction* (26): 58–83.

Elliott, Anthony. 2019. The culture of AI: Everyday life and the digital revolution.

Gao Jiaxi, and Haiyan Huang. 2023. "Stochastic optimization for energy economics and renewable sources management: A case study of solar energy in digital twin." *Solar Energy* 262: 111865.

Ghoreishi, Malahat, and Ari Happonen. 2020. "Key enablers for deploying artificial intelligence for circular economy embracing sustainable product design: Three case studies." *AIP Conference Proceedings* 2233(1): 840–855.

Grieves, Michael W. 2023. "Digital Twins: Past, Present, and Future. "In *The Digital Twin*, Cham: Springer International Publishing. In: Crespi, N., Drobot, A.T., Minerva, R. (eds) 97–121.

Hatuka, Tali, and Hadas Zur. 2020. "From smart cities to smart social urbanism: A framework for shaping the socio-technological ecosystems in cities." *Telematics and Informatics* 55: 101430.

Huang, Jiahui, Salmiza Saleh, and Yufei Liu. 2021. "A review on artificial intelligence in education." *Academic Journal of Interdisciplinary Studies* 10: 206–217.

Huang, Pei-hua, Ki-hun Kim, and Maartje Schermer. 2022. "Ethical issues of digital twins for personalized health care service: Preliminary mapping study." *Journal of Medical Internet Research* 24(1): e33081.

Israilidis, John, Kayode Odusanya, and Muhammad Usman Mazhar. 2021. "Exploring knowledge management perspectives in smart city research: A review and future research agenda." *International Journal of Information Management* 56: 101989.

Jafari, Mina, Abdollah Kavousi-Fard, Tao Chen, and Mazaher Karimi. 2023. "A review on digital twin technology in smart grid, transportation system and smart city: Challenges and future." *IEEE Access* 11: 17471–17484.

Jiang, Haifan, Shengfeng Qin, Jianlin Fu, Jian Zhang, and Guofu Ding. 2021. "How to model and implement connections between physical and virtual models for digital twin application." *Journal of Manufacturing Systems* 58: 36–51.

Jiang, Ji Chu, Burak Kantarci, Sema Oktug, and Tolga Soyata. 2020. "Federated learning in smart city sensing: Challenges and opportunities." *Sensors* 20(21): 6230.

Karimi, Yaghoob, Mostafa Haghi Kashani, Mohammad Akbari, and Ebrahim Mahdipour. 2021. "Leveraging big data in smart cities: A systematic review." *Concurrency and Computation: Practice and Experience* 33(21): e6379.

Kaut, Rohit, Chinedu Ossai, Abdur Rahim Mohammad Forkan, Prem Prakash Jayaraman, John Zelcer, Stephen Vaughan, and Nilmini Wickramasinghe. 2023. "The role of AI for developing digital twins in healthcare: The case of cancer care." *Wiley Interdisciplinary Reviews: Data Mining and Knowledge Discovery* 13(1): e1480.

Khan, Huma H., Muhammad N. Malik, Raheel Zafar, Feybi A. Goni, Abdoulmohammad G. Chofreh, Jiří J. Klemeš, and Youseef Alotaibi. 2020. "Challenges for sustainable smart city development: A conceptual framework." *Sustainable Development* 28(5): 1507–1518.

Lee, DonHee, and Seong No Yoon. 2021. "Application of artificial intelligence-based technologies in the healthcare industry: Opportunities and challenges." *International Journal of Environmental Research and Public Health* 18(1): 271.

Lee, Min Hwa, Jin Hyo Joseph Yun, Andreas Pyka, Dong Kyu Won, Fumio Kodama, Giovanni Schiuma, Hang Sik Park. 2018. "How to respond to the fourth industrial revolution, or the second information technology revolution? Dynamic new combinations between technology, market, and society through open innovation." *Journal of Open Innovation: Technology, Market, and Complexity* 4(3): 21.

Liu, Yi, Chao Yang, Li Jiang, Shengli Xie, and Yan Zhang. 2019. "Intelligent edge computing for IoT-based energy management in smart cities." *IEEE Network* 33(2): 111–117.

Luusua, Aale, Johanna Ylipulli, Marcus Foth, and Alessandro Aurigi. 2023. "Urban AI: understanding the emerging role of artificial intelligence in smart cities." *AI & Society* 38(3): 1039–1044.

Mashaly, Maggie. 2021. "Connecting the twins: A review on digital twin technology & its networking requirements." *Procedia Computer Science* 184: 299–305.

Mora, Luca, Roberto Bolici, and Mark Deakin. 2017. "The first two decades of smart-city research: A bibliometric analysis." *Journal of Urban Technology* 24(1): 3–27.

Nkwunonwo, Ugonna C., Felister E. Dibia, and Joseph A. Okosun. 2023. "A review of the pathways, opportunities, challenges and utility of geospatial infrastructure for smart city in Nigeria." *Geo Journal* 88(1): 583–593.

Park, Eunil, Angel P. Del Pobil, and Sang Jib Kwon. 2018. "The role of internet of things (IoT) in smart cities: Technology roadmap-oriented approaches." *Sustainability* 10(5): 1388.

Qu, Juncong, Mehmet S. Kizil, Mohsen Yahyaei, and Peter F. Knights. 2023. "Digital twins in the minerals industry–a comprehensive review." *Mining Technology* 132(4): 267–289.

Sadri, Habib, Ibrahim Yitmen, Lavinia Chiara Tagliabue, Florian Westphal, Algan Tezel, Afshin Taheri, and Goran Sibenik. 2023. "Integration of blockchain and digital twins in the smart built environment adopting disruptive technologies—A systematic review." *Sustainability* 15(4): 3713.

Schroeder, Greyce N., Charles Steinmetz, Ricardo N. Rodrigues, Achim Rettberg, and Carlos E. Pereira. 2021. "Digital twin connectivity topologies." *IFAC-Papers OnLine* 54(1): 737–742.

Shahat, Ehab, Chang T. Hyun, and Chunho Yeom. 2021. "City digital twin potentials: A review and research agenda." *Sustainability* 13(6): 3386.

Surya, Batara, Firman Menne, Hernita Sabhan, Seri Suriani, Herminawaty Abubakar, and Muhammad Idris. 2021. "Economic growth, increasing productivity of SMEs, and open innovation." *Journal of Open Innovation: Technology, Market, and Complexity* 7(1): 20.

Surya, Riza Afita, and Rif'atul Fikriya. 2021. "History education to encourage nationalism interest towards young people amidst globalization." *Waskita: Jurnal Pendidikan Nilai dan Pembangunan Karakter* 5(1): 1–13.

Telo, Joan. 2023. "Smart city security threats and countermeasures in the context of emerging technologies." *International Journal of Intelligent Automation and Computing* 6(1): 31–45.

Trindade, Evelin Priscila, Marcus Phoebe Farias Hinnig, Eduardo Moreira da Costa, Jamile Sabatini Marques, Rogério Cid Bastos, and Tan Yigitcanlar. 2017. "Sustainable development of smart cities: A systematic review of the literature." *Journal of Open Innovation: Technology, Market, and Complexity* 3(3): 1–14.

Voigt, Isabel, Hernan Inojosa, Anja Dillenseger, Rocco Haase, Katja Akgün, and Tjalf Ziemssen. 2021. "Digital twins for multiple sclerosis." *Frontiers in Immunology* 12: 669811.

Wagg, D. J., W. Keith, R. J. Barthorpe, and P. Gardner. 2020. "Digital twins: State-of-the-art and future directions for modeling and simulation in engineering dynamics applications." *ASCE-ASME Journal of Risk and Uncertainty in Engineering Systems* 6(3): 030901.

Winkowska, Justyna, Danuta Szpilko, and Sonja Pejić. 2019. "Smart city concept in the light of the literature review." *Engineering Management in Production and Services* 11(2): 70–86.

Wirtz, Bernd W., and Wilhelm M. Müller. 2023. "An integrative collaborative ecosystem for smart cities—A framework for organizational governance." *International Journal of Public Administration* 46(7): 499–518.

Xia, Haishan, Zishuo Liu, Maria Efremochkina, Xiaotong Liu, and Chunxiang Lin. 2022. "Study on city digital twin technologies for sustainable smart city design: A review and bibliometric analysis of geographic information system and building information modeling integration." *Sustainable Cities and Society* 84: 104009.

Zheng, Yaxin, Harm van Essen, and Regina Bernhaupt. 2023. "Utilizing service design approach to apply digital twins in home automation." Proceedings http://ceur-ws.org ISSN 1613: 0073.

Index

Pages in *italics* refer to figures and in **bold** refers to tables.

"

For Product Safety Concerns and Information please contact our EU
representative GPSR@taylorandfrancis.com
Taylor & Francis Verlag GmbH, Kaufingerstraße 24, 80331 München, Germany